PRACTICAL HANDBOOK
OF SOLID STATE
TROUBLESHOOTING

Other Books by the Author:

Workbench Guide to Electronic Troubleshooting
Practical Handbook of Low-Cost Electronic Test Equipment
Manual of Electronic Servicing Tests and Measurements

PRACTICAL HANDBOOK
OF SOLID STATE
TROUBLESHOOTING

Robert C. Genn, Jr.

Illustrated by E. L. Genn

Parker Publishing Company, Inc.

West Nyack, New York

© 1981 *by*

Parker Publishing Company, Inc.
West Nyack, New York

Library of Congress Cataloging in Publication Data

Genn, Robert C
 Practical handbook of solid state troubleshooting.

 Includes index.
 1. Electronic apparatus and appliances—
Maintenance and repair. 2. Semiconductors—Mainte-
nance and repair. I. Title.
TK7870.2.G45   621.3815′028′8   80-23897
ISBN 0-13-691303-2

 Printed in the United States of America

To Laura Chick, who held it all together
for so long

THE KEYNOTE FOR THIS BOOK ...

... lies in one word: Practicality. You will find it is not only extraordinarily practical, but comprehensive as well. Here are just a few of the new solid state servicing techniques you'll find in this book: state-of-the-art digital troubleshooting, up-to-date stereo system servicing techniques, modern solid state color TV troubleshooting, testing OP AMPS, and a multitude of other time-saving guidelines on solid state circuits. No important factor in modern-day electronics servicing is left out or passed over without crystal clear coverage ... all supplemented and further clarified by precise illustrations that pinpoint special things to bear in mind. These complete explanations will help you understand the *very latest* solid state devices, modern troubleshooting techniques, and just about everything else about solid state servicing that is of vital concern to today's serviceman.

Naturally, we'll include guidelines for servicing equipment containing the latest IC's along with scores of invaluable tips and shortcuts for testing and troubleshooting these revolutionary devices. For example, even if you have no background in mathematical logic or basic programming concepts, you can service these electronics circuits using nothing but the suggested test gear and information in Chapters 8 and 10. In those chapters, you will learn about the binary system, truth tables (and how they can help your troubleshooting), flip-flop circuits, registers and counters, adding and subtracting circuits, and much more up-to-date information that will help simplify every digital troubleshooting job you tackle.

Sooner or later, you will need concise, up-to-date guidelines on how to service the new amplifiers now being used in modern stereo systems. Chapter 1 includes actual circuit configurations and waveforms related to some of the newest techniques, as well as the theory of operation of these *super-high-fidelity* audio amplifiers. You'll find this chapter is written with emphasis on *simplified* servicing of solid state audio circuits, aided by the many detailed illustrations and practical servicing procedures that are streamlined for on-the-job application.

In addition, Chapters 2 and 5 describe servicing techniques that will speed up symptom analysis for linear IC's—differential and operational amplifiers, for instance. Each section contains time-saving troubleshooting procedures plus information that is of prime importance for understanding and servicing circuits containing these devices; procedures that you can instantly use on the job.

These are only a few of the modern servicing techniques and time-saving troubleshooting methods you'll find in the book. You'll discover there are literally hundreds, encompassing everything from SCR's right up to the most sophisticated IC system. Each section offers crisp, simplified troubleshooting procedures for handling modern solid state problems with a minimum of time and effort, providing all the information you need in order to understand and accomplish more in this important area of electronics.

Robert C. Genn, Jr.

CONTENTS

9

PRACTICAL HANDBOOK
OF SOLID STATE
TROUBLESHOOTING

CHAPTER 1

Troubleshooting Modern Solid State
Audio Amplifiers

How to Minimize System Troubleshooting Time

Cutting audio troubleshooting time when working with solid state stereophonic and quadraphonic systems, or any other transistor audio equipment, begins with developing a trouble-shooting procedure that is practical, fast, and dependable. The "secret," assuming there is one, to finding out what's wrong with solid state equipment is twofold: one, a thorough knowledge of how the circuit under test works—in other words, understanding the basic circuit you're troubleshooting—and, two, a swift way to move directly toward an accurate diagnosis of the problem at hand. So let's start out with a time-saving, troubleshooting technique that has been streamlined for instant on-the-job use, before going on to the actual procedures you'll need for the specific amplifiers you'll be working on. Basically, you only need to perform five steps to pinpoint the defective circuit—best of all, this technique requires practially no test equipment.

Step 1. **Analyze the system.** For instance, if you're checking a hi-fi set, check to see which particular feature does not work right. Operate the equipment and decide whether the trouble lies in one of the channels or somewhere else in the system. Check each component; the tape or cassette deck, the turntable, AM-FM radio, etc. This will cut your troubleshooting time because it quickly narrows the problem down to a particular section.

Step 2. **Make an examination** of the section that operates incorrectly. As an illustration, if a section of the set, or external component, is completely "dead," check its individual power supply. Another trick is to switch the input from one stereo channel to the other to see if it will bring the non-operating channel back to life.

Step 3. **Determine which stage is defective.** If you're not familiar with the equipment, a block diagram or circuit diagram should be consulted and each of the symptoms you found in Steps 1 and 2 considered.

To localize the defective stage (for instance, a transistor and its associated components), your first step is to make a visual check of all components and connections. Next, if the entire section is one IC or a plug-in module, and if you have a spare or a set available that uses the same components, try substitution. Your tests will depend on the type circuits you're checking (transistor or IC), and what the circuit is designed to do. However, at this point your tests can usually be made with nothing but a volt-ohm-millimeter (VOM) and, the main thing, you're still saving time and effort because you're narrowing the problems down to a single stage.

Step 4. **Isolate the faulty circuit** within the stage. The best way to isolate a bad circuit is to use the signal-tracing technique. This method requires some form of signal source and usually an instrument to measure the output signal of the circuit you're testing (this method is described in other sections of this chapter).

If you are out in the field and short of equipment, one way to isolate a faulty circuit (especially if the trouble is an intermittent), is to use a technique often called "Sledge Hammer Modulation" (in other words, gently strike or thump the equipment cabinet). I wouldn't suggest you do this in front of a customer, and you may find it causes a few raised eyebrows among some unknowing electronics technicians. But, believe me, many times it works. Your next best step is to start off *tapping* through the circuits *very lightly*, using the eraser end of a pencil. This technique can expose quite a number of troubles such as bad solder joints, cracks in PC board conductors, and dirty plug-in contacts. Still another way is to open the supply circuit and insert an ammeter. The current drain will give you an idea whether the circuit is acting as a short or open.

Step 5. **Localize the component** that is operating incorrectly or not at all. Set your VOM to check the transistor collector voltage and look for any transistor that is completely on (you'll read a lower-than-normal collector voltage), or completely off (you'll read a higher-than-normal voltage more than likely very close to the supply voltage). Either one of these voltage readings is an indication of a defective component. Once you've completed these five steps, it shouldn't take much time to pinpoint the component at fault and, with these troubleshooting techniques, you can usually uncover a defective **IC** as easily as any other components.

Understanding and Troubleshooting
Pulse-Width Modulated Amplifiers

During the past several years we have been hearing more and more about pulse-width modulated amplifiers, often referred to as "Class D" or "Switching Amplifiers." Since the beginning, the dream of amplifier designers has been greater effeciency, smaller size and less weight, and it appears that Class D amplification or, more properly, pulse-width modulation (PWM) is the answer. By the way, sometimes you may hear these amplifiers called "digital." However, it should be clearly understood that there is a distinct difference between digital and pulse-width modulation. For example, digital modulation usually involves a process of quantitizing the voice, music, etc. In one method of doing this, the instantaneous value of the analog signal (that is, the signal before it is digital) is sampled at specified intervals and the sample is converted into a digital number. Referring to Figure 1-1, you'll see the letters a, b, c, and d on the right-hand side of the diagram. The general idea of the process follows, starting with a. (See page 20.)

a.   Audio waveform (made unidirectional).

b.   Below b is the exact value, to four significant figures, of each signal point at variously spaced sampling instants shown at the end of each dashed line.

c.   Below c are the signal values, changed to nearest whole value.

d.   Below d are the values converted to binary form. These numbers, transmitted in sequence, will form the digital signal.

In one typical Class D amplifier, with no input signal you'll find that the output signal is a simple ultrasonic frequency (500 kHz) square wave. This serves as a carrier for the program information. Actually, the modulation can be done in several ways. For example, a few that you may encounter are; pulse-amplitude modulation, pulse-frequency modulation, pulse-position modulation, and pulse-width modulation. The last one is one of the easiest to achieve and is being used in some of today's hi-fi equipment. Figure 1-2 shows these

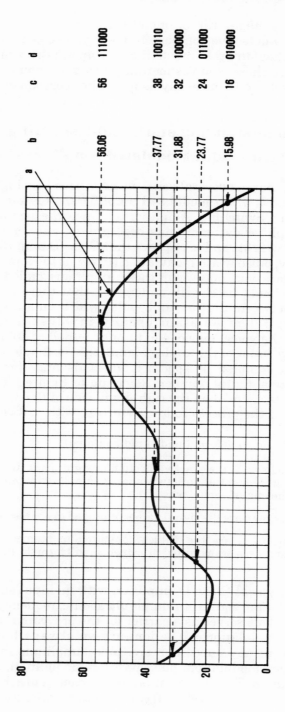

Figure 1-1:   The analog to digital conversion technique shown is often
referred to as "Quantitizing."

four different basic forms of pulse modulation. It should be pointed out that there are linear **IC's** on the market that are capable of producing time delays. For example, Signetic's NE/SE556 that may be used for both pulse-width and pulse-position modulation.

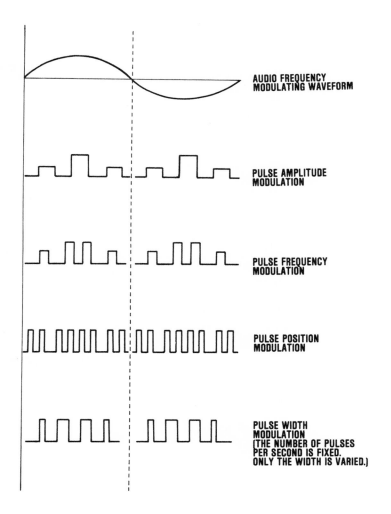

AUDIO FREQUENCY
MODULATING WAVEFORM

PULSE AMPLITUDE
MODULATION

PULSE FREQUENCY
MODULATION

PULSE POSITION
MODULATION

PULSE WIDTH
MODULATION
(THE NUMBER OF PULSES
PER SECOND IS FIXED.
ONLY THE WIDTH IS VARIED.)

Figure 1-2:   Basic pulse-modulation techniques

You'll notice that the PWM technique has a number of pulses-per-second that are *fixed* and it's only the process of varying the width of successive pulses that conveys the audio waveshape. A block diagram of how this may be done in a modern PWM hi-fi amplifier is shown in Figure 1-3.

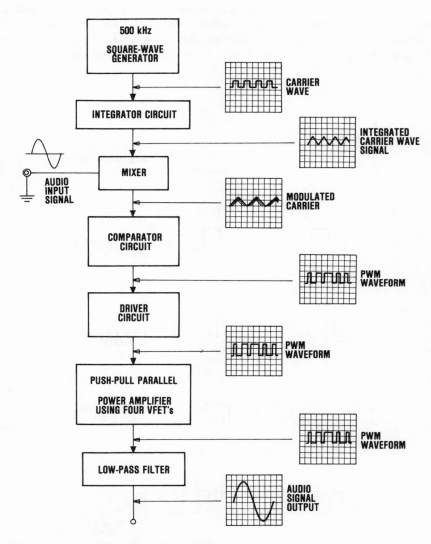

Figure 1-3:   Simplified block diagram and circuit waveforms of a pulse-width amplifier used in a modern hi-fi-system

When troubleshooting these circuits, starting at the top of the diagram shown in Figure 1-3, you'll see a 500 kHz square-wave generator common to both channels in the hi-fi amplifier system we are using as an example. Next, the waveform is converted to a sawtooth by integrator circuits and then run through a mixer, where it is added linearly with the incoming audio signal. The comparator circuit has two input signals; the modulated carrier wave and a crystal-controlled reference voltage. The composite signal out of the mixer is continuously compared to this reference voltage. The composite signal is then converted to a PWM square-wave signal and fed into a driver stage that feeds the final power output stage. Since both N- and P-channel FET's are available, the driver circuits for the vertical field-effect transistors' (**VFET**) power amplifiers are fairly simple and do not require a phase inverter stage. Finally, the PWM signal is passed through a low pass filter where it is converted to a sine wave and fed to the speakers. Sony designed their low-pass filter to have a flat frequency response up to 40,000 Hz when the amplifiers are terminated with 8-ohm speakers, which is why these hi-fi systems are sometimes called "Super-High-Fidelity."

Another state-of-the-art feature that is designed into these hi-fi ammplifiers is a switching regulated power supply. Because of their great efficiency, switching regulators are much better than the older types in packaging considerations as considerable saving in volume and weight can be attained. This becomes increasingly true when designing for high-output power levels (for example, 160 watts per channel in Sony's amplifier). You'll find switching regulators are discussed in Chapter 3.

Speaker Impedance Measurement, Matching and Compatibility

When we refer to speaker impedance, it must be remembered that we are talking about the *AC resistance* (i.e., the vector sum of reactance and resistance) seen looking into a speaker or combination of speakers. The true impedance cannot be measured with an ohmmeter, but must be observed using an AC signal. For modern speakers, the manufacturer's rate their speakers' impedance using a frequency usually located at about 400 Hz. Figure 1-4 will give you some idea of the complexity of today's ceramic magnet woofers and why most technicians don't try to repair them. (In case you didn't know, a woofer is a large-cone-area speaker that reproduces low frequencies.)

Impedance becomes very important when you use more than one set of speakers. Most stereo receivers have terminals for two sets of speakers, set 1 and 2, and possibly a switch to select set 1 or 2, or 1 + 2. In general, most stereo amplifiers won't drive more than one pair of 4-ohm speakers *without overheating*. What this boils down to is that you should use 8-ohm speakers for both pairs, if you are going to play set 1 + 2. On the other hand, as we've said, most amplifiers will overheat if you use more than one set of 4-ohm speakers. So, if you

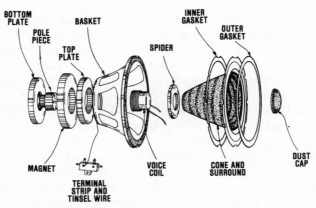

Figure 1-4:   Exploded view of a ceramic magnet woofer

want maximum power and are going to use a single set of speakers, 4-ohm speakers are best.

You'll find that transistorized amplifiers will usually work with loads (speakers or resistors) in the 4 to 16-ohm range. Of course, they will work a little harder driving the 4-ohm loads—draw more power at a given volume control setting than 8-ohm speakers—however, you'll rarely have trouble using the 4-ohm loads. *Caution:* **Two 4-ohm speakers connected in parallel result in a load of 2 ohms,** *not 4 ohms.* It's important to remember that although you *can* make changes in speaker 'hookups,' it is only true if you finish up with the original impedance—two 8-ohm speakers in series substituted for one 16-ohm, etc.

If you are installing a speaker system that requires long runs of speaker wire (10 feet or so), do not use No. 22 gauge wire that is commonly sold as speaker wire by many retail stores. It's best to use two-conductor stranded copper wire (zipcord) such as ordinary lamp cord (which is No. 18), to prevent excessive resistance between the amplifier and speaker. A good rule of thumb is: when in doubt, use a smaller gauge wire. Don't forget, the larger the gauge number of a piece of wire, the smaller the wire. Table 1-1 shows the maximum one-way length and gauge of copper wire you should use between an amplifier and a specific speaker.

After you have the speakers hooked up to a stereo set, energize the system and put it in a play, record, or tape mode, or tune to an AM or FM station. Next, increase the volume until you hear sound coming out of the speakers. If you don't hear any sound, use the following checklist.

1.   Check the AC power plug.
2.   Check the equipment ON switch.

ONE-WAY LENGTH	4-OHM LOAD	8-OHM LOAD
UP TO 18 FEET	GAUGE NO. 18 OR 20	GAUGE NO. 20
UP TO 30 FEET	GAUGE NO. 16 OR 18	GAUGE NO. 18 OR 20
UP TO 48 FEET	GAUGE NO. 14 OR 16	GAUGE NO. 16 OR 18

Table 1-1: Recommended lengths and gauges of speaker connecting wire. Technically, the smaller gauge number is the best of the two gauges given.

3. If the tape or record is moving, check where the tuner switch is set (AM, FM, tape, or phono position).

4. Listen for inter-station, unrecorded tape, or blank record noise. Turn the volume up, if necessary.

5. Check the speaker wiring hookup. Do you have the wires connected to the correct terminal? Most stereos have many possible connections in the back. Speakers generally won't work if you connect them to the phono or other external equipment input.

6. Try changing the speaker select knob, if there is one.

7. Check all speakers for shorts. If you find one, it's possible that you may also have a blown fuse or popped circuit breaker.

8. Use your VOM and check the speaker wires for continuity.

9. Try switching or replacing a speaker. If you get sound on one channel or the other, it's an indication that one channel is dead. See the five-step method of troubleshooting audio amplifiers in the beginning of this chapter for your next steps.

How to Service Small Signal Solid State Audio Amplifiers

All statistics indicate that audio servicing is one of today's strong, profitable businesses and, in most cases, you deal with stereo sets where you'll find small signal audio amplifiers are very common. Figure 1-5 shows a schematic of a typical transistor small signal (a few hundred milliwatts output) audio amplifier.

Troubleshooting one of these small amplifier systems is very easy. First, check the DC voltage supply. Incidentally, if you're checking a record player, you'll find the design has, more than likely, used the secondary winding on the turntable motor as a step-down transformer in the power supply. But don't let this worry you because, other than this, the DC power supply usually is a very simple half-wave rectifier.

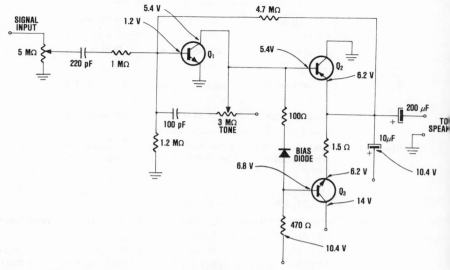

Figure 1-5:   Transistor small signal amplifier showing approximate
                   voltage readings that you should expect to find at the various
                   points shown

Next, referring to Figure 1-5, measure the voltage at the
emitter of Q3. You'll notice that you should read about 6 volts, in this
case. You may ask, "Why start at Q3?" The reason is that starting at
this point lets you check several things. If you read zero volts on the
emitter, it's a good chance that Q3 is open. Also, check the base
voltage of Q3. If it measures a very high negative voltage, the
transistor is cut off. This could be caused by the bias diode, or it could
be a problem with the transistor Q1, its circuit, components, and
supply voltage.

Let's say that you read a voltage just slightly higher or lower
than the one shown on the emitter of Q3. In this case, suspect a bad
Q1. It's probably a leaky transistor and should be checked in a
transistor tester. However, before you pull Q1, check the voltage
reading on the collector. If it's quite a bit off the listed voltage (about
5V, in this case), you can be pretty sure that your problem is that Q1
is leaky and affecting Q2 and Q3.

Generally speaking, you can use this troubleshooting
procedure on just about any small audio amplifier system. But there
are variations! For example, it is common to encounter *Darlington
amplifiers* in these systems. *A few words of warning:* If you remove a
Darlington, you must replace it with a Darlington or two equal
transistor amplifiers. During your troubleshooting, *be careful*
because they have three leads and look just like any other bipolar
transistor.

Another low-power solid state audio amplifier that you'll frequently encounter is one that is designed around integrated circuits (**IC's**). Figure 1-6 is an illustration of a stereo preamplifier designed by Fairchild, using a μA 739 **IC**. This preamplifier is particularly suitable for stereo applications where a low-noise, high-gain pre-amp is desired. You'll find it used in stereo tape inputs and similar applications.

The basic test procedures for audio amplifiers designed around **IC's** are essentially the same as for transistor audio amplifiers (see the first section of this chapter). That is, we must determine which stage fails to process its signal correctly. In fact, pinpoint the stage that messes up or blocks the signal, and you're more than halfway to eliminating the problem.

Two troubleshooting procedures have been used for a long time; signal tracing and signal injection. Both are very practical, both are fast, both are extremely dependable when you're sure of the signal being fed into the stereo, etc. As an illustration of how simple the signal injection technique can be, suppose you have a stereo amplifier like the one shown in Figure 1-6 and channel B has no output signal but channel A does. You're probably thinking "That's easy. Switch the A input signal over to B, and if you get an output, it's an indication the trouble is before channel B's input." Of course this is correct, but sometimes it's so easy that we overlook the obvious answer and, therefore, it bears repeating.

A single stereo amplifier like the one shown in Figure 1-6 isn't difficult to check because only two things can go wrong with the **IC**. It can develop an internal short that you can quickly detect with a DMM because almost every DC voltage will have an improper value, or, on the other hand, the **IC** can mess up signals and all DC voltages will check out all right. In this case a scope can help pin down the problem. If you happen to have a stereo sweep generator, it's much easier to produce a superior job. However, a sweep generator isn't really needed for fast troubleshooting and analysis. You can use a radio receiver signal, output from a turntable, or tape recorder, any of which will serve as a signal during preliminary troubleshooting. Whatever you use for an input signal, start out by turning everything on, then give the instruments a few minutes to warm up. Next, clip the scope test probe to one of the amplifiers. If you're using a signal generator, set it to the manufacturer's recommended input level, operating at the recommended frequency (use 1,000 Hz, if you don't know what the manufacturer recommends and set your signal level just below overdriving the stage). You'll see flattening of the output waveform on the scope, if the circuit is overdriven.

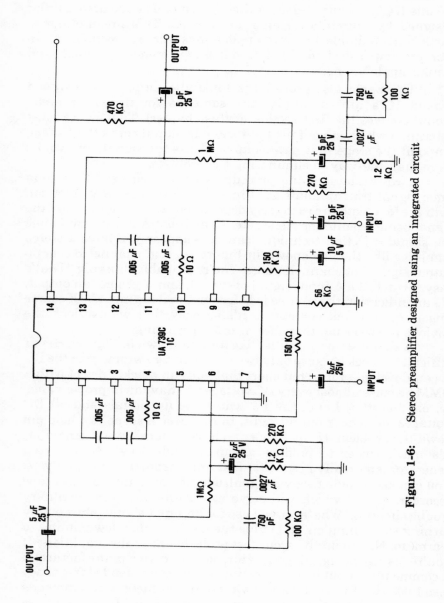

Figure 1-6:   Stereo preamplifier designed using an integrated circuit

If the amplifier you're troubleshooting is provided with any operating and adjustment controls (volume, treble, bass, etc.), these controls should be checked to be sure that all controls have the same effect across the bandwidth of the amplifier. During the check, don't forget that the treble control should have the greatest effect at the high end of the bandpass, while the bass control will have the most pronounced effect in the low end. *Note:* be sure to monitor the output of your signal generator because, in almost every case, it will change with a change in frequency. This can cause considerable error in your readings, so be careful. It is recommended that you use a constant output signal level and reset it each time you change your test frequency.

Troubleshooting Medium-Power Solid State Audio Amplifiers

Figure 1-7 is a schematic diagram of a medium-power amplifier that includes an integrated circuit, the Signetics NE 540. Power capacity is 35 watts, which is medium compared to today's solid state amplifiers that can easily exceed 100 watts. (Schematic diagram courtesy of Signetics, Menlo Park, California.)

Troubleshooting this type amplifier follows the same general methods outlined in the beginning of this chapter. Use the five-step method: Analyze the *system,* examine the malfunctioning *section,* determine which *stage* is defective, isolate the faulty *circuit* in the stage, and, finally, localize the bad *component.* With an input signal of any kind (even a small transistor radio's output signal), and an 8-ohm speaker connected to the output of the schematic shown in Figure 1-7, you can follow the signal through a simple power amplifier such as this one. Of course, a scope is best but almost any inexpensive VOM will tell you if the DC voltages are incorrect. In general, you'll find defective transistors will be your most common problem. So check transistors first. You'll find that in-circuit transistor checkers are very informative if you use them correctly. However, if you find a short, it's best to remove the transistor from the circuit for an absolute check before making your decision that the transistor is definitely in need of replacement.

In the begining of this chapter, quite a bit was said about speakers as an amplifier load. You can apply power to many solid state audio amplifiers with the speakers disconnected (*there are exceptions).* But under no circumstances should you *short* the speaker terminals of a solid state amplifier unless you are *positive* the circuit has overload protection. Another thing to watch for is when the speaker is connected across one-half of an output transformer. In

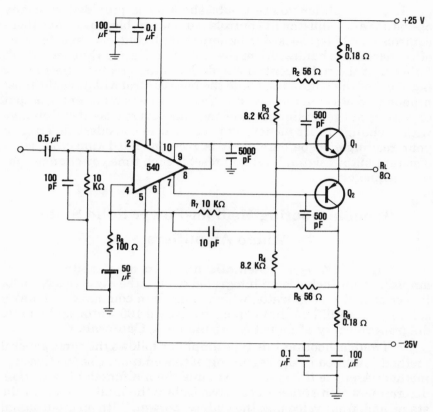

Figure 1-7:   An example of a Signetics NE 540 IC used as a power
amplifier. You will usually find an amplifier such as this
mounted on a PC board. Also, a 35 watt IC *must* be mounted
on a heat sink during operation.

this case, *never* operate the equipment without a suitable load such
as a resistor or speaker. Also, because we are now talking about
medium-power solid state audio amplifiers (up to about 100 watts),
the power rating of the speaker you use must be taken into
consideration. Incidentally on some speakers, you may not be able to
tell which terminal goes to positive and which goes to ground. *It
doesn't make any difference as long as both speakers are connected
the same way.* Why should they be connected the same way?
Because if they aren't, the speakers will be out of phase, resulting in
a weak bass and poor stereo effects. In most cases, you'll find a
phasing mark on one of the speaker terminals. This may be a dot of
paint, a + sign, or, if they aren't marked, you can simply watch that
both speaker cones move in and out *together*.

Guidelines for Servicing High-Power
Audio Amplifiers

You can troubleshoot high-power amplifiers (70 watts or more per channel in stereo systems) using the same methods described for the smaller amplifiers in the earlier sections of this chapter. The major differences are the number of components you'll need to check when working with them, and the different voltage readings you'll

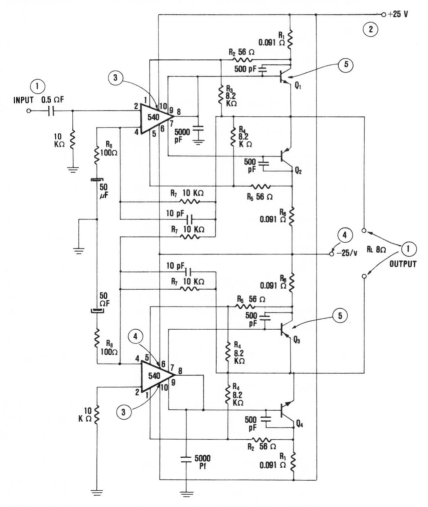

Figure 1-8:   Preliminary troubleshooting tests needed to isolate a bad stage in a stereo system

encounter. *Note:* any medium or high-power **IC** *must* be mounted on a heat sink *before making any kind of operational checks.*

To begin, check all channels to see if any one of them will pass a signal using all possible signal inputs, i.e., tape deck, phono, radio, etc. If you can't get anything on a certain stage, check the DC voltages, find the component, and check for shorts or possible leakage. Figure 1-8 (Courtesy of Signetics, Menlo Park, California) shows some of the measuring points that should be checked during your troubleshooting. The schematic could be one channel of a 140 watt stereo system. The point marked 1 on the left side of the diagram requires some form of signal input such as a signal generator, and the point marked 1 on the right side of the diagram requires a terminating load (8 ohms, in this case). All other points are DC measurements.

Key Steps to Troubleshooting Output Stages and Their Loads

Perhaps the most frequently used output stage in transistor audio equipment is the "Output Transformerless" (**OTL**, for short) amplifier. You'll find this circuit used in everything from the smallest to the largest audio power amplifiers. A couple of other terms you'll hear when working with **OTL's** are **stacked circuit** and **complementary amplifiers**. Figure 1-8 shows a schematic of the complementary output stage configuration.

Referring to Figure 1-9, you'll see there are some similar features. Namely, both the complementary and stacked circuits use two transistors with the load impedance connected directly to the junction or center-tap between two transistors. However, it's important to note that in Figure 1-8 two of the transistors are PNP's (Q2, Q3), and the others are NPN's (Q1, Q4). This is the first clue that the circuit is a complementary amplifier. Why? Because a complementary amplifier is an amplifier that utilizes the complementary symmetry of NPN and PNP transistors.

Now, let's take a look at Figure 1-9. Notice that the two transistors shown (Q2, Q3), are the same type (NPN). Does this circuit have any advantages over the other? Well, for one thing, notice that the circuit shown in Figure 1-8 requires a +25 volts and the −25 volts from the power supply, whereas the stacked circuit requires only a single voltage. Another thing that's important to know during troubleshooting these two (or similar) circuits, is that the one shown in Figure 1-8 usually is biased at Class AB. In other words, you'll always find at least some current flowing through the transistors. On the other hand, the circuit in Figure 1-9 usually is

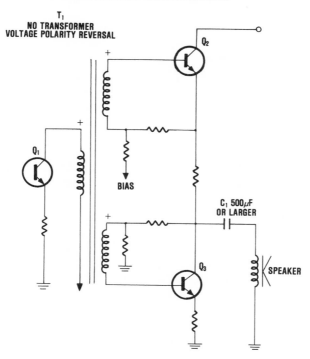

Figure 1-9:   A popular OTL circuit. Notice it uses a single-ended DC power supply; the clue is that one of the output transistors always goes to ground. Also, notice the large coupling capacitor (C_1). You'll find such a capacitor in almost every type of OTL circuit.

biased at Class B (all current is cut off except when you inject a signal).

How does the circuit shown in Figure 1-9 work? First, notice that the transistor Q_2 has no DC supply current when Q_3 is cut off. Next question. How can this be a linear amplifier circuit if both transistors are cut off during part of the input sine wave? The answer is that the transistor Q_3 is not really cut off during operation since the large capacitor C is discharging during this period. In fact, when current flows through Q_2, it charges capacitor C_1, and when Q_2 cuts off, C_1 discharges through Q_3 and provides the current necessary for the second half of the input waveform. What do we have now? A Class B amplifier with all its efficiency (about 70%).

Some valuable clues when troubleshooting are (1), use your VOM and check the DC power supply voltage. If the voltage is normal (not too high and not too low), both transistors are good. (2) Check the midpoint (between Q_2 and Q_3) voltage. If you find no voltage, either Q_2 or C_1 may be your trouble. In the case of distortion, you can use your scope to determine just where the defect lies. In most audio amplifiers, you'll find feedback loops. Signal-trace these feedback paths and you can pin the trouble spot down very quickly. Incidentally, these feedback loops are used for the same reason they are in any other amplifier—to improve the linearity.

Referring to the schematics we've shown for **OTL** amplifiers, you'll see that the speaker is all-important as far as amplifier load impedance is concerned. In other words, the speaker impedance is the only load these amplifiers have. This means that under no circumstance should you try using any speaker as a replacement except one having the correct impedance. If you decide to redesign the speaker system used with an **OTL,** just remember the formula:

power delivered to the speaker(P_L) = current through the speaker squared (I^2) times the impedance of the speaker (R_L)

or $P_L = I^2\ R_L$

This important formula can save you some peculiar symptoms. For example, a speaker turned inside out, maximum smoke and/or loud screams if you destroyed an expensive speaker.

Simplified Testing of FET's and Other Transistors

Practical testing of modern solid state devices requires that you be able to identify component packages as well as schematic symbols. Figure 1-10 shows some of the most frequently encountered packages and a few typical Junction Field Effect Transistor (**JFET**)

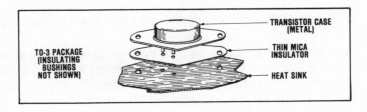

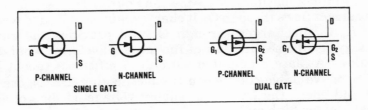

Figure 1-10: Some typical case configurations and schematic symbols that will be encountered in solid state troubleshooting

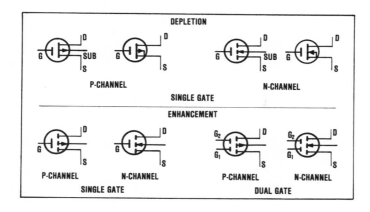

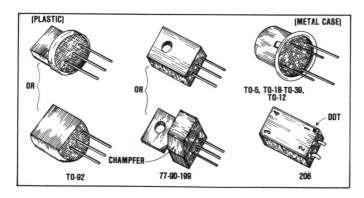

Figure 1-10 *(continued)*

and Metal Oxide Semiconductor Field Effect Transistor (**MOSFET**) schematic symbols that are used throughout the industry today.

Modern audio circuits use complementary - symmetry and stacked (totempole or quasi - complementary - symmetry) output stages in many cases. Basically, these stages are all alike, as has been explained in the preceding sections of this chapter. Of course, there are variations but troubleshooting them is about the same as any other solid state circuit. There may be larger or smaller transistors, diodes, etc., but when testing them, you can use a transistor tester (or VOM, not so reliable) for a quick check. When you are making a leakage test, remove the transistors from the circuit. Also, it is common for large germanium transistors to have a *permissible* leakage current of several hundred microamperes. On the other hand, silicon power transistors shouldn't have any leakage that can be detected on a low-cost instrument such as a VOM, etc.

When using a VOM to test a transistor, diode, or similar device, normally you will get a very low resistance reading both ways (example, emitter-to-base, base-to-emitter) if the device is shorted. This is especially true when checking silicon transistors. In fact, you'll probably read zero ohms, if it's shorted. One thing on your side when troubleshooting power transistors is that, generally, when a power transistor fails, it really fails!—a completely open or a dead short, with the short being the most common trouble.

Ohmmeter tests can be made on **FET's**, with some degree of success. Figure 1-11 shows the resistance readings you should expect when checking **JFET's.** The **MOSFET**, sometimes called *Insulated Gate FET (IGFET),* also can be tested out-of-circuit, using an ohmmeter. *Warning:* since it is easy to damage some **MOSFET's** with an ohmmeter, it is not recommended that you switch ohmmeter leads back and forth without some form of protection. The easiest way to protect unknown or easy-to-damage MOS devices is to connect a 1-megohm resistor (any wattage will do) in series with your ohmmeter lead, when measuring from drain-to-gate or source-to-gate. You probably won't have the same resistance readings as listed here, but if you can read less resistance one way and higher resistance the other, the transistor is probably all right. However, when in doubt, remember the ultimate check of any solid state (or tube, for that matter) device, is substitution of a known-to-be-good component. *Be careful* when using a standard VOM because it is possible that the instrument's voltages can permanently damage the small electrolytic capacitors used in much of today's electronic equipment. Many are rated at a working voltage of only 3 volts. Figure 1-12 shows the resistance readings you should find when checking a good **MOSFET** out-of-circuit.

Referring to Figure 1-11, you'll see that the drain-to-source ohmmeter reading should be the same regardless of how you connect your ohmmeter. You may read any value from about 100 ohms to 10k ohms. It doesn't matter, just so you get the same reading both ways.

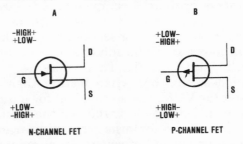

Figure 1-11:   Ohmmeter readings that should be expected on a good out-of-circuit JFET. "A" is an N-channel, and "B" is a P-channel type.

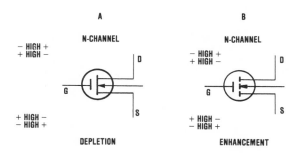

Figure 1-12: Ohmmeter reading that should be expected on a good out-of-circuit MOSFET. "A" is a depletion/enhancement mode and "B" is an enhancement mode.

Next, the gate-to-drain should be identical, depending on how you connect your ohmmeter. If you connect the ohmmeter leads in the forward direction (low resistance), you'll read somewhere around 1 k ohm. Reverse the leads (high resistance direction), and you should see a reading of infinity (open circuit).

Figure 1-12 shows that you should see an open circuit reading on the ohmmeter regardless of how you connect the ohmmeter between the gate and other leads. When checking the other possible connections, you'll probably find a drain-to-source resistance of about 1 k ohm in the forward direction (schematic A). The other, (B), should read between 100 and 10 k ohms in a low resistance direction and show as open in the reverse, as you can see. Typical case diagrams for the **FET's** are shown in Figure 1-13.

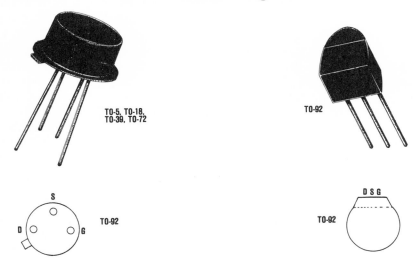

Figure 1-13: Common packages used for N-channel and P-channel FET's

Replacing Bipolar Transistors

Sometimes the difficult part of servicing solid state equipment is finding a suitable substitute for a defective transistor. There are several replacement lists that can be a real lifesaver when the going gets rough. One is to get a solid state manufacturer's substitution guide or a list of transistors that are designed to replace almost any transistor you happen to encounter. Another source that I've found to be very helpful is the transistor substitution guide for replacement transistors written by Robert and Elizabeth Scott and published from time-to-time by the magazine, *Radio Electronics*.

Of course, there's always the transistor that pops up occasionally that can't be found on any available list. Even these aren't always hard to replace. To begin, when you're working with audio equipment, you don't have to worry about the replacement transistors cut-off frequency, which helps considerably. There are really only a few simple things we need to know in the beginning of our search. These are:

1. What type transistor is needed (NPN, PNP) and is it a germanium or silicon?
2. What is the minimum break-down voltage that you can tolerate between the collector and emitter?
3. What is the minimum current gain (h_{fe} or beta) of the transistor you're to replace?
4. What kind of package should the replacement transistor have? This one isn't too important because, as a last resort, you probably can modify.

Let's say you find that you must replace an NPN, silicon transistor. It has the same type case as the transistor package shown as TO-92 in Figure 1-10. The breakdown voltage between collector and emitter (V_{CEO}) has to be at least 25 volts. The first question is easily solved with an ohmmeter. The type package requires nothing but looking at the faulty transistor. Next, and probably the most important, is the breakdown voltage. Check to see what peak-to-peak signal voltage level is normally applied to the defective transistor. Whatever it is, simply double it. Your answer is the minimum breakdown voltage you should have. If you don't know, or can't determine the signal voltage level, measure the DC voltage level on the defective transistor leads and use these as a last resort.

Now, with all this information, your next step is to find a transistor that has *approximately* the same characteristics. Several manufacturers and sales outlets publish transistor manuals and spec sheets that list ratings and characteristics. For example, Radio Shack and General Electric Co., are just two of many.

Assume that you decide a 2N3394 will fit your needs. Your next step is to refer to a transistor substitution handbook or list. Turning to the transistor number 2N3394, you'll find there are *many* transistor substitutes for this one. I came up with 23 of them. However, I have to admit I picked an easy one to use as an example.

CHAPTER **2**

Guidelines for Servicing Today's Integrated Circuits

As you know, **IC's** are among the most common components used in electronics. They come in all sizes, shapes, classifications, and you must select, identify, check, remove and install them. **IC's** and circuits designed using **IC's** generally are easy to troubleshoot and understand once you get the hang of it. This chapter is full of practical time-saving techniques for working with all kinds of **IC** circuits, using nothing but conventional test equipment.

You'll find that the emphasis has been placed on *work-related problems*. For example, selecting a replacement **IC,** linear **IC** amplifier distortion test, constructing and/or using audio test loads, troubleshooting **IC** bandpass amplifiers, measuring dynamic-output-impedance, and how to make a "safe" output power measurement, are just a few of the situations that you'll find in the following pages.

Most of us in the electronics service field favor practical ideas over ideology and never before has technical "know-how" been so well rewarded as it is today. Every section in this chapter provides you with the practical troubleshooting techniques that are essential to your increasing success as a technician.

Essentials of Modern IC Troubleshooting

Integrated circuits are classified in a number of ways. The most useful classifications for the troubleshooter are:

(1) **Method of Construction:** Although how an **IC** is constructed is not of prime importance to the troubleshooter, he should have a basic understanding of terms used in the manufacturer's technical literature. For example, you will usually find that the method of construction identification is described in the "spec sheets" as *monolithic, thin film,* or *hybrid.* One way to describe a monolithic **IC** (abbreviated **MIC**) is to say that it is a complete electronic circuit fabricated on a single chip that cannot be penetrated without destroying it. The most important point is that the monolithic **IC** is the most common type you'll encounter when servicing modern electronic equipment. You'll see why as you read on.

Basically, a *thin film* **IC** is a very simple circuit and usually consists of passive elements such as capacitors, resistors and conductors. Normally, you won't find active components like transistors and diodes built into this type **IC** (probably the most important point to make regarding troubleshooting them). By the way, you'll also encounter thick-film methods used to construct the same type circuits as thin film. Although a different process is used during their manufacture, there isn't any difference in trouble-shooting either of the devices.

The next one, the *hybrid* **IC** (**HIC**) includes both the monolithic and film types just described (see Figure 2-1 for package illustrations). The advantage, if there is one, of the hybrid, in reference to the monolithic, is that transistors, diodes and passive elements may be used most efficiently. This advantage is offset by the low reliability of the hybrid and is one reason why the **MIC** is the type **IC** found in most of today's electronic equipment.

(2) **Operating Mode:** The operating mode we are most interested in, in this chapter, is the *linear* (the *digital* mode is discussed in Chapters 8 and 10). Linear operation of **IC's** is the same as in any other electronic device; any signal modification (such as amplification or detection) must be accomplished without amplitude or other distortion. Furthermore, in most cases, the input signal is sinusoidal. However, just because a particular **IC** was originally presented by the manufacturer as a non-linear **IC** does not mean it can't be used in a linear mode. In fact, this is frequently done ... particularly by electronics experimenters. A great many manu-facturers supply Applications Memos that suggest using their **IC's** both ways. For example, Signetics say their 531 monolithic operational amplifier may be used as a voltage-controlled amplifier (linear operation), or as either a half-wave or full-wave precision rectifier (non-linear). As another example, it is common to find that a hi-fi circuit experimenter has used a multiple input digital **IC** as a linear mixer amplifier. Furthermore, linear **ICs** such as the

differential and logarithmic amplifiers are not always operated in a linear fashion.

When looking through the different manufacturers' catalogs and Application notes, you'll find many linear **IC's**: all types of amplifiers, analog voltage comparators, complete communications circuits (such as AM radios and video circuits), operational amplifiers, Darlington amplifiers and so forth.

(3) **Type Packaging Used:** Sometimes, because of space requirements, this is important when you're servicing. There are other times when space isn't important but exact type replacement is. Comparing the packages will usually help, in this instance. Figure 2-1 will give you some idea of the packages used for **IC's**. But, it must be emphasized that the ones shown are only a small sample of the total number used. Although I haven't counted them, I'd guess at least a hundred different packages have been used by the various manufacturers.

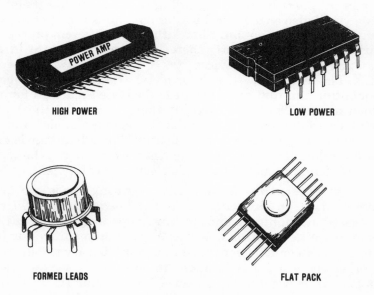

Figure 2-1: Typical IC packages that you may encounter during troubleshooting

Of course, when looking at the schematics of the equipment you're troubleshooting, you won't find drawings of the **IC** packages as shown in Figure 2-1 ... that comes later, when you start to examine the actual circuits. When checking a schematic, more than likely you'll find a symbol such as the one shown in Figure 2-2 (A). However, if you're referring to a manufacturer's catalog, service

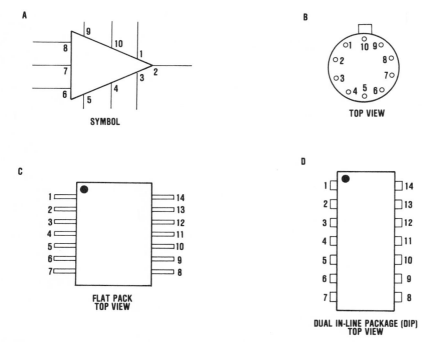

Figure 2-2:   Typical **IC** pin configurations

notes, or spec sheet, you'll probably find the pin configuration as shown in Figure 2-2(B), (C) or (D).

Interpreting Current Data Sheets

When troubleshooting or debugging systems using **IC's**, data sheets are a tremendous help, provided you can read them. When using data sheet information, or similar catalog data, it usually is necessary to interpret several pages of graphs, etc. The problem is that most **IC's** are not single components. **IC** electrical data is for a complete electronic circuit. Therefore, the performance of an **IC** cannot be described by some single parameter (such as beta of a transistor or the mutual conductance of a tube), but by a series of graphs and ratings that describe the operation of the particular **IC**.

To begin our discussion, let's use the 540 monolithic, Class AB power amplifier shown in Figures 1-7 and 1-8, as an example of how to interpret the information published by the manufacturer. Although each company has its own system of data sheets, and it's impractical to try to discuss them individually, the following information and terms are typical of what most provide.

Open-loop gain and frequency response: Open-loop gain usually is defined as the ratio of a change in output voltage to a change in input voltage of a *loaded* amplifier *without* feedback. The open-loop gain is both frequency and load (R_L) dependent, as shown in Figure 2-3. This particular graph is for the Signetics power driver 540. It is specifically designed to drive a pair of complementary output transistors, as shown in Figures 1-7 and 1-8. However, you may encounter similar graphs that show how the supply voltage affects gain in reference to frequency. These illustrations are called *open-loop voltage gain graphs* and are used to show the performance of operational amplifiers (**OP AMP's**). Generally, gain increases with supply voltage and decreases with increased frequency. The effects of temperature are often included, but a change in temperature will usually produce a change in gain everytime there is a change in frequency. On the other hand, if you hold the temperature constant on an **IC** such as the 540, the open-loop gain will decrease slightly as the temperature increases. For example, Signetic's graph of the 540 shows a gain of 93 at –50°C and a gain of about 89.5 at +125°C.

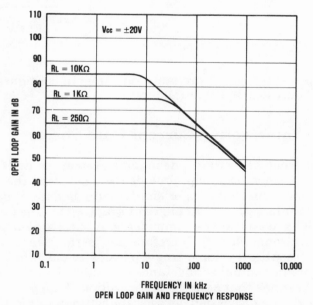

Figure 2-3:   Open-loop gain and frequency response for Signetic's power driver 540. Courtesy Signetics, Menlo Park, California

Closed-loop frequency response: A closed-loop usually describes a circuit in which the output is continuously fed back to the input for constant comparison. Therefore, in almost every case, the manufacturer will include a schematic of the circuit used when recording the data for the graph. As an example, see Figure 2-4 for a closed-loop frequency response graph and test circuit for the 540 **IC**.

These graphs can be used directly to select values of feedback components as well as for excellent troubleshooting aids and replacement guides. Once you understand the data provided, it becomes very apparent that, in most cases, you have quite a large tolerance when selecting replacement components (capacitors, resistors, etc.), regardless of the type **IC** you're working with.

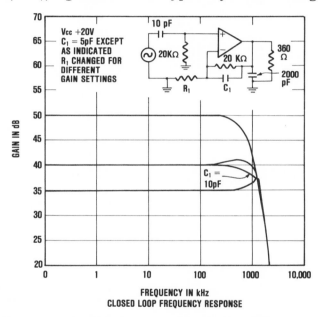

Figure 2-4:   Closed-loop frequency response for a Signetics power driver 540. Courtesy Signetics, Menlo Park, California

Phase response versus frequency: Figure 2-5 shows a phase response curve for a Signetics power driver 540 (sometimes such curves are included on other frequency response graphs). In any circuit, there will be some phase shift between input and output signals. Although this usually is not critical for audio amplifier circuits, there are exceptions such as **IC** operational amplifiers, where feedback is used to control gain, etc. In these circuits, the closed-loop gain is considered the worse case of phase shift because a closed-loop circuit allows a gain of unity, permitting the highest frequency response.

The following list defines other characteristics that you'll find either in graphical or tabular form in **IC** manufacturer's literature.

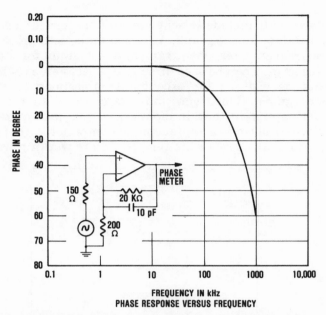

FREQUENCY IN kHz
PHASE RESPONSE VERSUS FREQUENCY

Figure 2-5:   Phase response versus frequency for a Signetics power driver
540. Courtesy Signetics, Menlo Park, California

Bias Current: Sometimes a graph of bias current versus
temperature. Normally, it is the current into the **IC** input terminals
measured during the application of a signal (ordinarily expressed in
microamperes or picoamperes).

Common-Mode Input: Figure 2-6 shows a differential
amplifier connected in a common-mode input configuration. The
gain formula is shown in Figure 2-6, and the measurement is made
with the signal applied in phase (i.e., common-mode) to both inputs
of the differential amplifier.

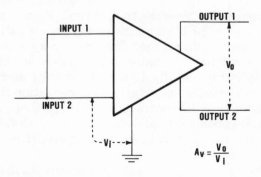

Figure 2-6:   Common-mode input (differential amplifier)

Common-Mode Input Impedance: The open-loop input impedance of both inverting and non-inverting inputs of an operational amplifier, with respect to ground, or the input impedance between either input of a differential amplifier **IC** and ground, and, in both cases, expressed in ohms. Figure 2-7 shows a typical linear **OP AMP** with inverting and non-inverting inputs indicated.

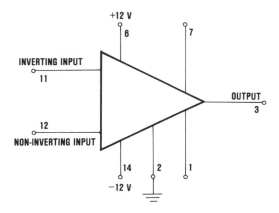

Figure 2-7:   Typical linear OP AMP showing inverting and non-inverting input terminals

Common Mode Rejection Ratio: Sometimes this parameter is listed simply as common-mode rejection or in-phase rejection. The manufacturer may be telling you how well the device (differential amplifier, in almost every case) performs with both inputs open and with both inputs common. It is the ratio of the gain when measured under these two conditions. Although it is usually stated as a dB equivalent of two voltage gain measurements taken at a specified input signal frequency, it should be pointed out that all manufacturers do not agree with this definition. For example, a widely used definition is: the ratio of the common-mode input voltage to the output voltage, expressed in dB.

In actual practice, the two inputs to the differential stage are balanced against each other so that when there is no input signal to both inputs, or equal input signals to both inputs, there should not be any output signal. An input signal to only one input of the device (see Figure 2-6), or an unbalance of two input signals, produces an output signal proportional to the difference between the two input signals. Figure 2-8 is a graph of the common-mode rejection ratio as a function of frequency (note the change of definition) for a differential video amplifier μA733. The 733 is a monolithic differential input, differential output, wide-band video amplifier.

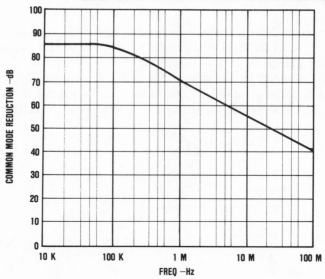

Figure 2-8:   Common-mode rejection ratio as a function of frequency.
Courtesy Signetics, Menlo Park, California

Differential-Input Impedance: This is the input impedance measured between the inverting and non-inverting input terminals of a differential input **IC** or operational amplifier (see Figure 2-7). Sometimes you won't find differential-input impedance listed in a manufacturer's specs. Instead, you'll find input resistance and input capacitance. However, in all cases, the measurements usually are made between the non-inverting and inverting input terminals of the differential amplifier. Like all other impedance measurements, the values are expressed in ohms. Also, input impedance will change with temperature and frequency, plus have an affect on loop-gain.

When troubleshooting **IC's**, it's important to remember that any alternations that affect input impedance will cause a change in amplifier gain. In practical terms, the **IC** input impedance is fixed. So, for maximum transfer of signal between the **IC** and its driving device, you should keep the two impedances fairly close to the same values. Therefore, if you're substituting either the **IC** or its driver circuits and can't tolerate any loss of signal strength, you must select components that will effect as near perfect a match as possible. This is definitely the exception to the rule in general troubleshooting of solid state circuitry because, in most cases, you can get away with a considerable mis-match without serious difficulties.

Differential-Input Voltage Range: You'll also find the term "input voltage range" used in **IC** spec sheets but there may be a difference between the two. Both are referring to the signal range that may be applied to the **IC** input terminals. For example, the differential video amplifier μA733 spec sheets list the *input voltage range* as ±1V. This **IC** uses pins 1 and 14 as the input signal pins. The manufacturer (Signetics) is saying that ±1V is a normal operating range. In the same spec sheets, under the heading "Absolute Maximum Ratings," the *differential-input voltage* is listed as ±5V. What this all means is that if you ever find a signal with a range of ±5V between pins 1 and 14 on this **IC**, you probably need a new one. But any voltage close to ±1V can be construed as good. Incidentally, the common mode maximum input voltage usually will be a little more (±6V, in this case).

Input-Offset Voltage and Current: Sometimes you'll find only the input-offset current listed and sometimes both will be shown on a single graph. But, manufacturers rarely show output-offset data without showing the gain and circuit connections used during the measurement. The reason for this is that the output voltage is set by the feedback.

The input-offset voltage may be used to inform you how much voltage (V_{in}) should be applied between the input terminals of a differential input with two equal resistors, to obtain a zero output voltage (V_{out}). The input-offset current is defined as the difference between the input bias currents (I_1 and I_2 in Figure 2-9), flowing into each of the inputs of the **IC**, when the output of the device is at zero volts. See Figure 2-9 for an example of test circuits used to measure the parameters of a differential video amplifier.

An **IC** amplifier data sheet usually lists several other characteristics. However, because **IC's** generally are designed for

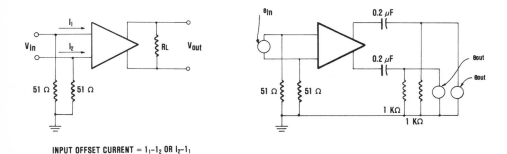

INPUT OFFSET CURRENT = $I_1 - I_2$ OR $I_2 - I_1$

Figure 2-9:   Test circuits used to measure the various parameters of a differential video amplifier. Courtesy Signetics, Menlo Park, California

specific applications, it is all but impossible to list every characteristic you could encounter. Furthermore, many of the characteristics aren't important for everyday troubleshooting. What we're most interested in is the typical information found on IC data sheets and how this data can be used by the troubleshooter.

IC Amplifier Input/Output Relationships

When selecting a replacement **IC**, or attempting to calculate circuit values, first you should determine if the **IC** can produce the rated voltage gain at the operating frequency. To do this, you can refer to data sheets similar to the one shown in Figure 2-3. Note that the gain should be about 84 dB's at 10 kHz, with a load (R_L) of 10 k ohms in the open-loop configuration. But you should expect a loss in gain using the same test frequency with a feedback circuit included. This is shown in Figure 2-4.

To calculate the output voltage (E_2), for any specific input voltage, probably the most important thing to note is that the gain in dB's lets you know what you should expect for an output voltage with a certain input voltage. However, it takes a little fancy mathematical footwork to figure it all out (of course, any inexpensive pocket calculator that does logs will simplify the problem).

To get an idea of how to solve such a problem, let's assume you have an **IC** that is rated for a gain of 40 dB at a certain frequency. To find the voltage gain of this amplifier, all you have to do is use the formula dB = 20 log E_2/E_1. Your work should look like this:

$$dB = 20 \log E_2/E_1 \quad \text{and} \quad 40 = 20 \log E_2/E_1$$

divide both sides of the equation by 20 and you get

$$2 = \log E_2/E_1$$

taking the antilog of 2 results in 100 = E_2/E_1

Now, what this means is that your output signal should be about 100 times larger than the input signal level (for example, 1 mV in, 100 mV out).

Effective Linear IC Amplifier Distortion Test

Although a dual-trace oscilloscope is more expensive, it is, by far the most effective and simplest to use for checking linear **IC** circuits for distortion. Distortion analysis is more effective using a square-wave signal generator as a signal source because of the high odd harmonic content. However, it should be pointed out that, in some cases, a sine-wave generator is better for testing than a square-wave generator (for example, when making power measurements or frequency response tests).

The procedure for troubleshooting, using either input signal, is essentially the same. You inject the input signal into the input of the circuit under test and monitor the output with a scope. The test setup for using a dual-trace (or single trace) scope, is shown in Figure 2-10. The value of the load impedance should be the **IC** amplifier's normal load impedance or the manufacturer's recommended test load.

Generally, it's bad practice to use a wire-wound load resistor because the internal reactance of the resistor may cause considerable error in your measurements. Although it isn't a problem when checking most **IC** amplifiers, there are times when output power, because of the voltage rating of the load resistor, may become troublesome. The trouble is in finding a carbon or composition resistor with a high enough wattage rating. Perhaps the best answer is to use a wire-wound resistor for testing high-power audio amplifiers, but you will experience slightly more reactance, thus introducing some error into your measurements. Incidentally, with the high-power stereo amplifiers we are servicing today ... 50, 100, 200 and even as much as 700 watts per channel ..., a solution for testing is to use an audio load such as those sold by Radio Shack, Heathkit, etc. These audio loads will handle up to 240 watts and are designed to connect to 2, 4, 8, 16, or 32 ohms outputs.

Regardless of what you use for a load, you are looking for any deviation of the output signal (shown as scope input A in Figure 2-10) from the input signal (shown as scope input B in Figure 2-10). If you observe any change in the square-wave input signal, the change

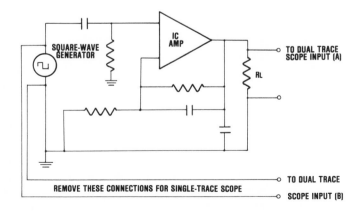

Figure 2-10:   Test setup for an IC amplifier distortion analysis

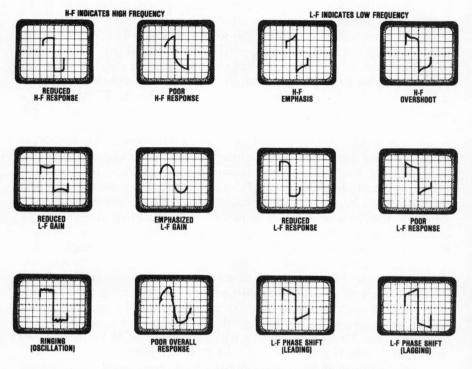

Figure 2-11: Oscilloscope patterns of a distorted square wave that could be observed during testing of linear solid state amplifiers

often can help pinpoint the problem. For instance, Figure 2-11 shows 14 distorted waveforms that you might see when looking at the input A trace on the scope.

When you make the check, it is best to use a high impedance probe on the scope. Check the value of the capacitor on the signal input lead of the amplifier, if possible, and then place a larger value capacitor between the signal generator and amplifier under test. This is to prevent distortion and maintain the safety of the amplifier under test. If you find you're having problems with low frequency response, check the input coupling capacitor and input resistor of the unit under test. The capacitor may be partially open, too small a value, or the resistor value may not be correct.

As another example, if you're having trouble with ringing (oscillation), it may be that distributed capacitance and inductance are at resonance at some low frequency. Try moving the external circuit leads to different positions. Sometimes it will help. If you are working with video equipment, it's possible that a peaking coil needs a damping resistor across it.

Checking an IC Bandpass Amplifier (RC Type)

Quite often, you'll find a bandpass amplifier in modern electronics equipment. One bandpass amplifier designed using an IC, uses a *notch filter* connected in the feedback loop of the IC. See Figure 2-12. As shown, the amplifier is quite narrow-band (has definite sharp peaking). However, with modifications, an amplifier such as this one may have a continuously variable Q (say from 20 Hz to 20 kHz). When modified, all the resistors and capacitors in the feedback loop usually are variable. Incidentally, the Q control in Figure 2-12 permits the bandwidth of the filter circuit to be varied over a limited range.

Troubleshooting these amplifiers is fairly simple because about the only thing that can go wrong is either the feedback circuit components change values or the IC goes bad (assuming the power supply is okay). The best way to start troubleshooting the circuit is to first check the supply voltages with a VOM. Typically, you should find about 14 volts DC on the IC power supply connection terminals. If the DC voltages check out all right, connect a scope at the output, as shown in Figure 2-12. Assuming you can get the circuit to operate with a signal input, your next step is to either compare your scope pattern with a known-to-be-good system of the same type, or refer to the manufacturer's service notes.

Two ways to locate a bad component are: (1) turn power off and use a VOM to check the resistors and capacitors (you'll have to lift one wire of each component to make each check because, as you can see, there are parallel circuits all over the place), or (2), connect a signal generator to the input and a scope to the output, as shown,

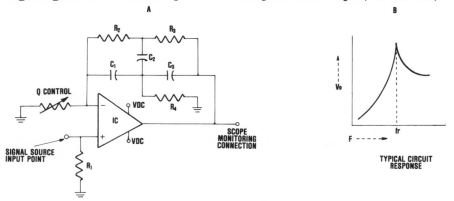

Figure 2-12: Simplified IC bandpass ampifier using a notch filter connected in the feedback loop. Connect test signal source and monitoring equipment to points shown in A. A typical frequency response is shown in B.

and substitute the IC, resistors, and capacitors until you find the troublesome one. Of course, there is another way. If you have the manufacturer's notes listing voltages, simply measure voltages until you locate the component with an incorrect voltage.

How to Make a Dynamic Load Test

It is important to realize that the *internal output impedance* of an amplifier is *not* the same as the manufacturer's specified load impedance. This is particularly true of negative feedback IC amplifiers. For example, you may encounter amplifiers using negative loop feedback designed to work into a 16-ohm load, although their internal output impedance may be only 1 ohm or less. In case you're wondering why, the reason for the difference is because such things as speakers continue to vibrate even after the exciting signal is removed. This will, of course, cause fuzziness, etc. when we listen to the music, records, or whatever. By making the internal output impedance of the amplifier lower than the load impedance, the designer builds in a damping effect that kills the unwanted vibrations. By the way, after reading this, it should be apparent that although we are often told we can get maximum power transfer by matching the source impedance to the load impedance ohm-for-ohm, it isn't always the best way to go.

It is common to use what is called *building out resistors* to effect an impedance match between the amplifier's internal impedance and rated load impedance. In general, the resistors are placed in series with the leads, between the amplifier and load. You'll find this done in many older systems such as the sound mixers used in radio stations, recording studios, etc. Generally, the most important thing we need to know pertaining to the load is what the value of impedance should be and how much power does the load have to handle. The next section will explain how to make a power output measurement.

Linear IC audio amplifiers of any design will be affected by any changes you make in the load. As you may recall from the previous chapter, most stereo amplifiers won't drive more than one pair of 4-ohm speakers without overheating. Therefore, it becomes apparent that a dynamic-output impedance measurement of an amplifier can be very important, especially if you don't know what load the amplifier is rated for, or suspect the output impedance has changed. The circuit for a dynamic-output impedance measurement is shown in Figure 2-13. Notice, a sine-wave signal generator is called for in this measurement. You may remember that it was emphasized that a sine-wave generator is better than a square-wave

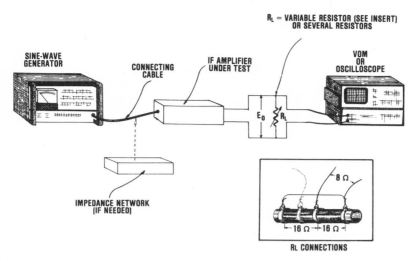

Figure 2-13: Test setup for making a dynamic-output impedance measurement. If a wire-wound resistor is used for R_L, it will introduce a slight error in your measurements.

generator for making certain tests (power response and frequency response measurements). We can now point out that if you want to check a power quantity using a sine-wave input/output with a resistance load, simply use the formula $P = E_o^2/R_L$ to determine average power (assuming you use an rms reading voltmeter, or convert the peak-to-peak voltage reading of a scope). Of course, you could use an ammeter and the formula $P = I^2 R$ or $P = IE_o$. One other point to keep in mind is that the IC has a power dissipation of its own. Therefore, *power out* of an IC amplifier is equal to *power in* minus *power dissipated*.

If you will examine the wiring connections of the load resistor in Figure 2-13, you will see there are two ways that the resistor can be connected in the circuit. You can connect to terminals (2) and (3) but, to achieve an 8-ohm output, terminal (2), i.e., the slider, would have to be set at one-quarter of the total resistance. Using this setup, the power handling capability of the resistor shown is only 25 watts. But, using the shorting wire between terminals (1) and (3), as shown, increases the power handling capacity of the resistor. In fact, wired in parallel, it can handle up to 50 watts when measuring the output impedance of an 8-ohm amplifier. Of course, checking a 4-ohm impedance using this system of resistor conections limits you to approximately 25 watts power handling capacity. But these connections are still better than connecting between the resistor terminals (2) and (3) without the shorting wire. By the way, using the same line of reasoning, it's possible to place another slider on the resistor and effectively connect three resistors in parallel.

Before applying power to the equipment setup as shown in Figure 2-13, set the load resistor to a high value—approximately 8 ohms is usually safe for solid state amplifiers. Energize the system, set any tone controls such as special bass boost, etc. to the OFF position and let the system warm up. With the generator set at 1,000 Hz, adjust the load resistor until maximum average power is found. De-energize the system and remove the load resistor from the circuit. Using an ohmmeter, measure the resistance. This value is the dynamic output impedance of the amplifier under test. Remember that the value you measure is valid only at the frequency of 1,000 Hz. A more detailed test can be made by repeating the test across the frequency range of the amplifier you are checking, using the frequency-by-frequency method or, easier, by using a sweep generator and scope.

How to Make a Maximum "Safe" Power
Output Measurement

The dynamic-output-impedance measurement just described can be used to calculate the power output of an amplifier circuit. But, to determine the *maximum safe output power*, it's easier to use a scope. This value is found by setting the signal level of the input signal to the amplifier under test (or volume level, if you're working on a hi-fi set, etc.) to a level where you just begin to see clipping on the scope and then reducing the input just slightly, i.e., just low enough so you no longer can see any clipping of the signal.

If you're in the field and don't have any equipment with you, a quick way to find a hi-fi system's maximum safe power output level is to use a tape of *loud* piano music as a test signal. Adjust the volume control of the set until you just begin to hear distortion. Mark the volume control at this point because this is where the amplifier is beginning to clip, because of overdriving. *Note:* this setting is not necessarily within the limits of the loudspeakers being used. Therefore, due to possible speaker damage, music should not be played under these conditions for long periods of time.

Servicing DC Voltage Amplifiers

There are several solid state analog or linear circuits used as DC voltage amplifiers. For example, there are many amplifiers designed around transistors that will accept and amplify DC voltage inputs. There are also hybrid circuits (similar to **IC's** in that resistors, capacitors, diodes, and transistors are all contained in a single package), that can function as DC amplifiers. One type **IC** that works well as a DC amplifier is the operational amplifier because it can be used with direct coupling and has good stability. Figure 2-14 shows a DC amplifier of this type.

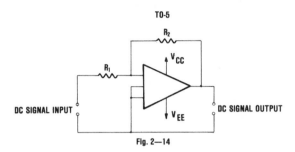

Fig. 2—14

Figure 2-14:   DC voltage amplifier of the IC operational amplifier type

If you don't have the manufacturer's service notes for the piece of equipment you are servicing, you can refer to the manufacturer's data sheets for the particular **IC** you're working with. For example, if the circuit shown in Figure 2-14 is found to be the stage where your trouble is located, just by eye-balling it we can see that the resistor R_2 is the feedback path of the **IC**. One clue is that R_2 usually will be about five times larger than R_1.

As you can see, checking a circuit such as this one is fairly simple. The first step is to measure the DC voltage at V_{CC}, V_{EE}, and the signal out. Assuming there is no output signal and V_{CC}, V_{EE} are good, remove the equipment source voltage and then check resistors R_1 and R_2 with an ohmmeter. If they both check out okay, it's a pretty safe bet that the trouble is a defective **IC**. The next best move is to try substituting another **IC** of the same type.

If you want to make a complete test of the circuit, the procedure is essentially the same as for any audio amplifier **IC** (see Chapter 1). However, keep in mind that the basic **IC** has a maximum input and output voltage limit, neither of which can be exceeded without possible damage to the **IC**. In general, the maximum rated input voltage should be applied and the output measured. Also, remember that the closed-loop gain is (or should be) dependent upon the feedback circuit resistance. Keeping these points in mind can help quite a bit during troubleshooting.

Single IC Radio Receiver Troubleshooting

A typical application of a linear integrated circuit is the AM radio receiver subsystem IC shown in Figure 2-15, courtesy of Signetics, Menlo Park, California. The NE546 is a monolithic IC that provides an RF amplifier, IF amplifier, mixer, oscillator, AGC detector, and voltage regulator in a single IC. It comes in a 14-lead dual in-line package and its primary application is in super-heterodyne AM radio receivers, particularly automobile receivers.

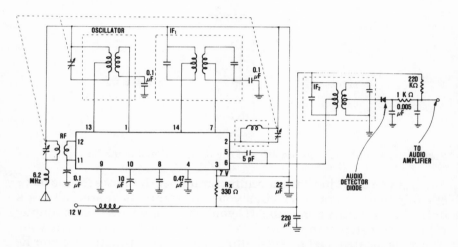

Figure 2-15:   Circuit diagram of a single IC capacitor-tuned AM radio

The absolute maximum ratings provided by the manufacturer for the NE546 IC are listed in Table 2-1 and are, as you can see, a great help when troubleshooting, with or without the equipment service notes. For instance, if you should measure a voltage of over 16 volts on pins 3, 13, 14, at pin 6, it's probable that the IC has been destroyed. As another example, more than 40 volts at pin 3 (DC supply voltage pin), could also produce maximum smoke and result in a charred IC.

Troubleshooting a circuit such as the one in Figure 2-15 requires voltages to be measured at radio frequencies; for example, the measurements made to the left of the audio detector diode at the right-hand side of the diagram, i.e., checking the two IF stages and oscillator sections shown in the dashed lines. When you measure voltages such as these and they are beyond the frequency capabilities of your meter circuits or scope amplifiers, an RF or demodulator probe is required. On the other hand, it's best to use a

ABSOLUTE MAXIMUM RATINGS	
V $_{CC}$ SUPPLY VOLTAGE PINS 3, 13, 14 AT PIN 6	16 V
DC SUPPLY VOLTAGE (V+)	40 V
DC SUPPLY CURRENT	35 mA
*INTERNAL POWER DISSIPATION	750 mW
LEAD TEMPERATURE	300° C
OPERATING TEMPERATURE RANGE	−40° C TO +85° C
STORAGE TEMPERATURE RANGE	−65° C TO +150° C
*RATING APPLIES FOR TEMPERATURES UP TO 55° C. DERATE LINEARLY AT 6.67 MW/° C ABOVE 55° C.	

Table 2-1:   Absolute maximum ratings for the NE546 IC

high-impedance probe to check the voltages after the audio detector. Typically, you'll find another **IC** being used in an audio amplifier circuit following the detector. Figure 1-8 is a schematic of such an amplifier and its test points (see page 31).

Troubleshooting IC Oscillators

Although you'll encounter many sine-wave **IC** oscillator circuits designed using LC-tuned circuits similar to the one shown in Figure 2-15, more recently an increasing number of oscillators (especially audio sine-wave generators) are being built without inductors. One of the reasons is that resistor-capacitor (RC) circuits are generally better at low frequencies than inductor-capacitor (LC) circuits, particularly in respect to size, weight, and, most important, *cost.*

There are several different RC-tuned networks used with **IC's**. All are quite old, starting way back in the days of tubes. They are the familiar RC phase-shift oscillator, RC twin-T oscillator, and the RC Wien-bridge oscillator. Figure 2-16 shows these basic circuits designed around an **IC** such as an OP AMP, etc.

Troubleshooting these AF oscillators is usually very simple. For example, when checking the components of this particular phase-shift oscillator (1), you'll probably find that all the resistors are of equal value in the phase-shift network (R_2, R_3, and R_4). Also, the capacitors C_1, C_2, and C_3 should be of equal value. If any of these components change value, it will cause the oscillator to either

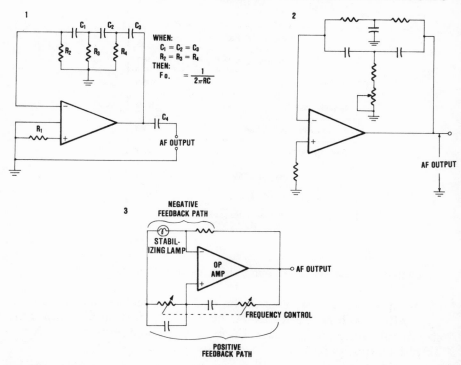

Figure 2-16:   Basic IC sine-wave audio oscillators using RC frequency generating networks

change frequency or stop oscillating, or both. If you find a decreased output signal level, it's very possible that one of the resistors has changed to some lower resistance value, or, the gain of the IC is insufficient. Refer to Chapter 6 for additional information on each of the oscillators discussed in this section.

     The twin-T oscillator (2) also uses a phase-shifting network from the output of the IC amplifier back to the input. The variable resistor must be properly adjusted to make the circuit oscillate. Also, as mentioned before, any change in value by either the capacitors or resistors in the phase-shifting network will result in a change of operating frequency. The approximate frequency of operation can be found by using the formula:

$$\text{frequency} = \frac{1}{2 \pi RC}$$

     The last AF oscillator (3) shown in Figure 2-16 is a Wien-bridge. For sine-wave AF oscillators, the twin-T oscillator just described and the Wien-bridge are used as often today as they were in the past.

     Like the other RC oscillators, the operating frequency of the

Wien-bridge is set by the value of R and C in the positive feedback path. When the two resistors are of equal value and the capacitors are matched (shown in the positive feedback path), the circuit will oscillate at a frequency determined by the formula just given for the twin-T. However, as with any oscillator, if the negative feedback is too great, the oscillator will cease to function. The lamp (Figure 2-16) is important to the operation of the circuit because it is used to regulate the amount of negative feedback.

When troubleshooting this circuit, be sure to remember that both the negative and positive feedback path components are important to proper operation. However, component values must be such that the positive feedback exceeds the negative feedback before oscillation can be sustained. Some experimenting with these circuits may be necessary to achieve stable oscillation; that is, you may need to adjust component values a little to make the circuit oscillate.

Servicing Touchplate IC Electronic Control Systems

Touchplate control circuits are appearing more and more in consumer products. For example, computer typewriter keyboards, microwave ovens, and TV tuning systems use touchplates. The actual construction of such circuits is greatly simplified through the use of **IC's**. A stripped-down version of one touchplate control system is shown in Figure 2-17.

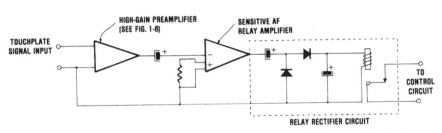

Figure 2-17: Simplified diagram of a touchplate control system. See text for servicing information

Troubleshooting these systems follows the same rules as for AF amplifiers outlined in Chapter 1. In some cases, an adjustment for sensitivity is included. The adjustment may be found installed either between the first and second **IC** amplifier or connected in one of the stages of either the amplifiers. However, for economic reasons, most systems do not include sensitivity adjustments.

The relay rectifier circuit enclosed by dashed lines in Figure 2-17, converts the AC output of the **IC** to DC to operate the relay. If you're experiencing difficulty with relay chatter or insufficient relay current (can't get the relay to close), check the electrolytic capacitors in this circuit—especially the output capacitor in older systems.

Assuming the system you're checking uses relays, if one won't actuate, yet the rest of the relays all perform normally, the trouble must be in the driver circuit or the relay itself. A simple ohmmeter check will tell you if it's the relay. If the relay checks out okay, go to the input of the relay driver **IC** and inject a signal. If this doesn't do the trick, replace the **IC**.

CHAPTER **3**

Successful Ways to Troubleshoot

Contemporary Solid State

Regulated Power Supplies

You'll find each section in this chapter is aimed at giving you a good practical knowledge of regulated DC power supplies found in all types of solid state equipment. These power supplies come in many shapes and sizes, from very small **IC's** up to monsters that have to deliver enough power to drive 700-watt audio amplifiers. This chapter will help you service them all.

The hi-fidelity amplifier, pulse-width modulated DC power supply serves as an excellent example of the type equipment you'll learn to troubleshoot. But, it should quickly be pointed out that you don't need new test gear to troubleshoot these sophisticated systems. Your regular radio-TV test equipment can be used to perform every job discussed.

This practical guide to troubleshooting modern DC power supplies includes reliable test procedures that you will use frequently on the job, important information about parts replacement, and answers to some puzzling symptoms found during troubleshooting. These are just a few of the *simplified techniques* and *shop hints* that deal with working with solid state regulated DC power supplies which are covered in the following pages.

Fundamentals of Modern Regulated
DC Power Supplies

It's all but impossible to speak of modern, regulated DC power supplies without bringing up the subject of *Pulse-Width Modulated* (PWM) power supplies. In the beginning, we should point out that there is some controversy about what these systems should be called. Some say "pulse-duration modulation" and others say "pulse-length modulation" or "pulse-time modulation," but whatever we call them, they all work on a principle in which the duration of a pulse is varied. Another term frequently used is *switching regulator*. However, they may or may not be pulse-width modulated.

Now that we've brought it up, you may be wondering "Why use a pulse and what's the purpose of changing the time duration of the pulse?" To answer these questions, let's first consider the difference in efficiency of the common single-ended Class A, B, and C amplifiers. You'll remember that a Class A amplifier is about 30% efficient, the Class B is about 50% efficient, and the Class C is 70% to 80% efficient. Next, you'll also remember that the Class A is operating 100 percent of the input signal duration period, Class B is operating 50 percent of the time, and Class C, less than one-half the time. From this, it appears that the more time we can keep an electronic device from operating, the more efficient it is. Which is true, provided the device can still perform the desired function within the imposed time limit. In other words, the main reason for using pulse-width modulation in a DC power supply regulator is that it uses less power than the older system and, therefore, we can achieve greater efficiency.

There are other advantages. When used in TV sets, the pulse rate normally is set at the horizontal sweep frequency, which permits the use of smaller (lighter) chokes and capacitors and still maintain the same filtering efficiency. Although they operate at a slightly different frequency (typicaly 20 kHz) than TV receivers, some hi-fi amplifiers use the same type power supply. Again, considerable space and weight are saved and efficiency is better, compared to a conventional 60 Hz power supply. Figure 3-1 is a basic block diagram of a PWM power supply.

Before we troubleshoot any electronic system, we should know how it works. Therefore, let's see how this circuit does its job. First, the bridge rectifier feeds its DC output to the controllable switch. This switch is off except for the time required to charge the filter capacitors in the filter circuit it feeds. Of course, like any other DC supply, the output of the filter circuit should be a steady, pure DC voltage. However, as we all know, this may not always be the case

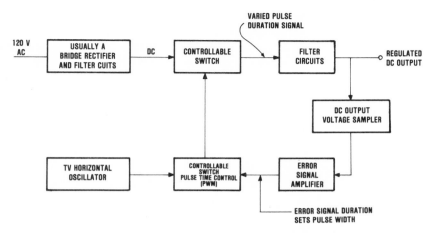

Figure 3-1:   Basic pulse-width modulated power supply diagram

(due to load variations), and that is the reason for the feedback loop back through the DC voltage output sampler, error signal amplifier, and on to the controllable switch pulse time control (pulse-width modulator). As you can see, the PWM stage varies the pulse duration. The prime purpose of varying the pulse duration is to maintain the filter capacitors at a certain DC voltage without utilizing any more energy than necessary. To put it another way, all the PWM does is recharge the filter capacitors at regular intervals, as the load on the supply discharges them, and it doesn't supply any more energy than necessary to charge the capacitor to a certain peak voltage.

What happens during actual operation is that when the load draws more current, the charge (energy) in the capacitors, decreases because more energy is being taken out than is being fed in. On the other hand, if the load current requirement decreases, the charge in the capacitor increases. Of course, the output voltage will rise and fall if we consider nothing but the voltage across the filter capacitors. But the error amplifier senses this voltage change and feeds back an error signal that is used to make corrections by adjusting the time the controlled switch is off and on, resulting in a constant, well-filtered output voltage.

You'll remember that when this type DC power supply is used in TV receivers, the pulse frequency used ordinarily is the horizontal sweep frequency (about 15,750 Hz). This should be kept in mind when you're servicing one of these TV sets because, if the horizontal oscillator isn't working properly, it's very possible that you won't have a DC output from the DC supply. More about this is in the next section on troubleshooting.

Now that we know that the operation of the DC supply depends on the horizontal oscillator, we can bring up the fact that all manufacturers include some form of starting circuit (usually two diodes, called *starting diodes*). The way the starting circuit works is that when the set is turned on, a small DC pulse is fed to the receiver's horizontal oscillator to start it operating. At almost the same instant, the output of the horizontal oscillator is feeding back to the DC supply to provide the operating pulses. When troubleshooting, it's important to remember that if the starting circuit is not working properly, you won't have a DC output from a PWM DC power supply.

In summary, a PWM power supply won't produce an output until the horizontal oscillator is operating and the horizontal oscillator can't work until the output stage of the PWM supply is working. Sounds impossible? It is, until you realize that all actions produce reactions and, in this case (provided all systems are "go"), *instantaneously*.

Troubleshooting Pulse-Width Modulated
DC Power Supplies

When troubleshooting switching regulator DC power supplies, your first step is to check the DC output voltage of the rectifiers. If you find the output voltage normal, your next step is to check the receiver's horizontal stages. An example of a rectifier circuit used by RCA is shown in Figure 3-2. *Note:* in many cases, all the DC voltages and waveforms of a PWM DC power supply *must* be measured in reference to the power supply's *isolated ground*. All chassis DC voltages *must* be measured in reference to *chassis ground*. Typically, on a schematic, isolated ground is indicated by a hollow triangle.

Let's asume that you find the horizontal stages are working properly. What's next? Well, as you'll recall, the starting diodes can also "kill" the whole power supply. To solve this problem, let's again use one of RCA's receiver power supplies as an example. You'll find the starting diodes in the power supply circuits that connect to the horizontal stages. Figure 3-3 is a simplified diagram of this section of the PWM power supply.

Before you make any active tests (apply test voltages), you should make passive tests. With your volt ohmmeter, check the starter diodes, filter capacitors, transformers, resistors, etc., for shorts and opens. Let's stop right here and say a few words about *fast-recovery diodes* before we proceed any further, because you might have to replace a diode at this point in your troubleshooting.

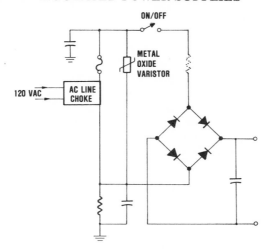

Figure 3-2:   Bridge rectifier used by RCA to provide unregulated B+ to a
TV receiver PWM power supply

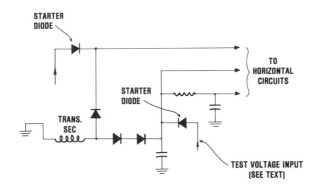

Figure 3-3:   An example of starter diode placement in an RCA color TV
receiver PWM DC power supply regulator

Most of the diodes we are talking about in these systems are of
this type, i.e., they have to turn off very rapidly. So, you should
always use a fast-recovery diode as a replacement in any kind of DC
power supply that operates at a high frequency (such as 20 kHz), or
at the horizontal frequency of a TV set. Incidentally, if the diode
quickly blows every time you put one in, your trouble probably is an
incorrect replacement. Now that the problem of diode replacement is
out of the way, let's get back to troubleshooting the PWM DC power
supply.

After you have assured yourself that there are no shorts or opens, RCA suggests that it may be possible to start the horizontal oscillator in the set we are discussing (chassis CTC-85), by applying a +22.5 DC pulse voltage to the +27 voltage input, as shown in Figure 3-3. To create a DC pulse voltage, use a 22.5 volt battery or other suitable DC power supply and simply make a quick make-and-break contact to the +27 volt input. You can use a DC voltage output somewhere near 22 volts. For example, a bias box (a variable bench-type power supply with a negative and positive voltage output) should work. Since all of today's PWM power supplies have starting circuits, this test should be valid for any of them. Of course, you will probably have to make some modification due to the different voltages and circuits used by manufacturers.

To check if the starter circuit is working, you can connect a voltmeter into the starter circuit and watch for an upward kick on the needle, when you connect the DC source. I've found that analog voltmeters are much better than digital voltmeters for making checks of sudden voltage changes, particularly if you're only looking for a short-time upward surge of current, or when checking capacitors, etc.

Almost any low-cost, modern oscilloscope can be used to check the control pulses and waveshapes in these power supplies, because their operating frequency is usually in the 17 kHz to 20 kHz range. It should be re-emphasized that the waveshapes are important to the correct operation of these supplies. Because of this fact, it follows that the scope is the best troubleshooting tool, even though you can use a VOM to make preliminary checks.

When you first see a schematic of one of these power supplies, it may cause you to utter a few harsh words. But these circuits aren't really hard to troubleshoot. Simply follow normal troubleshooting procedure, i.e., check to see that each stage is doing its job correctly and you'll have the set out of the shop in a surprisingly short time. To prevent getting the set back in the shop in an equally short time, here are a few words of caution. When replacing capacitors, chokes, transistors, and diodes, *be sure to use exact replacement parts*. The reason is that the voltages and currents these power supplies are handling are pulses operating at frequencies at, or near, 20 kHz. Ordinary electrolytic capacitors and other components just won't stand up under the high currents, at the frequencies they are subject to in these power supplies.

One other thing that should be mentioned is that although the bridge rectifier circuit shown in Figure 3-2 is standard today, it's very common to encounter special rectifier assemblies with four internal diodes internally connected with external connections. Figure 3-4 shows a full-wave bridge rectifier schematic diagram and

PACKAGE

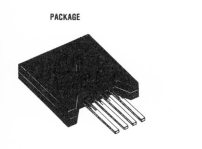

CIRCUIT

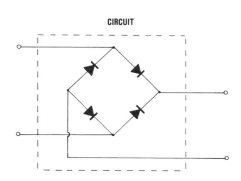

Figure 3-4:   Full-wave bridge rectifier schematic and actual package that
you may find used in electronics equipment

what the actual package looks like. Either positive or negative
output voltages usually can be obtained from these devices simply
by grounding the minus or positive terminal.

Understanding and Troubleshooting the
New Power Transformers

Look at the schematic of a pulse-width modulated DC power
supply and very likely, you'll find a ferrite core transformer similar
to the one shown in the circuit in Figure 3-5. We might ask what a
ferrite core transformer is and why it is used in place of the old
standby ... the conventional transformer. Well, to begin, a ferrite
core is a ceramic-like substance molded under high pressure and
composed of iron, nickel, zinc, manganese, and copper. The
compound is molded into the desired shape (there are many shapes
being produced for a host of different uses) and fired in a manner
similar to the ceramic art work you find in a local high school
ceramics class.

The transformer constructed around a ferrite core has
primary windings and secondary windings similar to a conventional transformer. But the input voltage to a PWM supply's
regulator ferrite core transformer commonly is around 20 kHz,
whereas a conventional transformer input voltage is usually 60 Hz.
What this means is that there would be considerable loss in the
tansformer core at this high operating frequency if the manufacturer did not use a special core material; in other words, a ferrite
core. Ferrite cores have extremely low eddy current losses and can be
magnetized and demagnetized very rapidly.

When troubleshooting this type transformer, you can check
the primary and secondary windings with an ohmmeter, just like

any other transformer. However, if you're making operational voltage checks, watch it! The input voltage is usually a rectangular wave with a frequency of around 20 kHz, which means a scope is your best bet during troubleshooting. Figure 3-5 is a basic block diagram of Sony's Class D amplifier's switching voltage regulator power supply, showing where the ferrite core transformer is located in the system and what type waveform you should see on your scope.

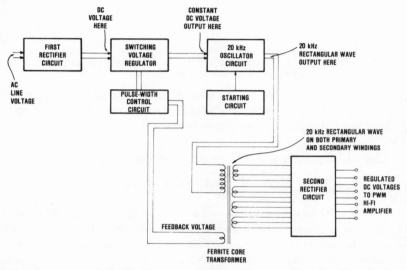

Figure 3-5:   Block diagram of a switching voltage regulator DC power
supply showing placement of the ferrite core transformer

    Although I haven't actually seen one, the Wanlass Electric Company is circulating information about a unique transformer called a *Paraformer*. Their information indicates that this new transformer would work well in a PWM power supply. The drawing in Figure 3-6 will give you an idea of the transformer's construction.

    It is possible that you might find one of these transformers used as a line voltage regulator or as a control in a switching type regulator. As a line voltage regulator, it would be placed between the first rectifier circuit (see Figure 3-2) and the 120 VAC line voltage source. On the other hand, this transformer could possibly be found as a replacement for the ferrite core transformer shown in Figure 3-4. Which procedure you use to troubleshoot the device will depend on where it is located. However, other than frequency and waveshape considerations, it would seem that standard troubleshooting techniques already explained could be used. One difference that you'll notice in Figure 3-6 is the capacitor connected across the

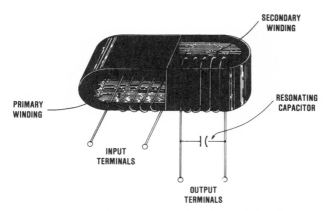

Figure 3-6: Basic drawing of a Paraformer transformer distributed by the Wanlass Electric Company

output leads of the transformer secondary. This capacitor makes the secondary a resonant tank circuit. While troubleshooting, it's best to think of this transformer as an oscillator operating at a frequency somewhere in the 10 to 20 kHz range. As in any other tank circuit, if the capacitor develops a short or open, everything grinds to a halt.

Compared to a conventional 60 Hz power supply, all the switching type transformers (operating with a high frequency rectangular wave input) and power supplies are far lighter and smaller. As with PWM amplifiers, these power supplies must be carefully shielded and filtered to prevent harmonics of the switching frequency from leaking out and getting into other equipment. This is called *electromagnetic interference* (EMI). For this reason, it is important that you replace all shielding ground connections, and filters, after servicing this equipment, or you'll probably experience EMI problems. Incidentally, one of the advantages of the Paraformer is that it normally does not need an rf interference filter in the main power lines because, due to its design, it cancels almost all transients coming in.

Testing Today's Output Capacitors

In a full-wave bridge rectifier circuit (or a standard half-wave, for that matter), if you measure a low output voltage, it's possible that the trouble is low capacitance or an open filter capacitor. As we all know, these capacitors act as a reservoir to hold the charge, therefore, the DC voltage must drop if there is a reduction in capacitance. The very simplest way to check one of these capacitors is to bridge another capacitor across the one you suspect. If the DC voltage returns to normal, your problem is solved. *Caution:* do *not* shunt large electrolytic capacitors across other in-circuit capacitors in solid state power supplies, with power on. If you do this, it may cause a large surge of current and can damage the equipment. *Always turn the power off before connecting capacitors into a circuit.*

If the output voltage does not increase when you connect an exact replacement capacitor in the circuit, the next step is to check the rectifier diodes, since a bad rectifier diode will produce the same symptom. When in doubt (especially when checking for open diodes), use your scope. An open rectifier diode will cause a pronounced increase in the ripple voltage. However, a *shorted diode* in a bridge ordinarily will reduce the output voltage by at least half. It also will probably blow a fuse or trip the circuit breaker.

Another point pertaining to the very large by-pass capacitors used in today's power supplies is that they should produce zero impedance (for all practical purposes) to ground at their operating frequency. *Note: ground may be an isolated ground; be sure to check this before making measurements.* If there is a high internal impedance, the signal return currents may develop voltages across the impedance. These voltage variations will then be fed back into the other stages where they can cause a multitude of problems.

If you're selecting replacement capacitors for a bridge rectifier system, the capacitors used should be electrolytics because, as has been pointed out, high values of capacitance are required. The voltage rating of the capacitors should be *at least* 1.3 times the DC output voltage. If you don't know the value of capacitance needed, or are designing a solid state bridge rectifier, you can calculate the value of capacitance (in microfarads) needed by using the formula:

$$C_{\mu F} \approx (2 \times 10^5)/(\text{load resistance} \times \text{maximum ripple permitted})$$

where, $\approx$ is approximately equal.

For example, let's assume you're working on a circuit that has a load resistance of 50 ohms (if you don't know the load resistance, just remember that in a *pure resistance load*, $R = E/I$). Next, the desired maximum ripple voltage is 5%. The capacitance you would want to use is:

$$2 \times 10^5 / 50 \times 5 = 800 \mu F$$

You'll notice that as the maximum ripple frequency permitted and load resistance are decreased in one or both their values, the value of capacitance needed is increased. Naturally, since we want the supply to be stable, a large output capacitor is the easiest way to insure a safe gain-phase relationship.

The best way to test filter capacitors is to simply touch the capacitor leads with your scope probe and look for any voltage variations. If you see anything except a straight DC line with the scope vertical gain set to maximum, it's wrong! For example, seeing a 60 or 120 Hz waveform (it may look like a sine wave or be distorted to look like a sawtooth wave), indicates the filter capacitor is low in value or, possibly, open. This test is valid for all the capacitors except the input capacitor. At this point, you'll usually find about 10 volts peak-to-peak. Of course, it must be remembered that we are talking about the bridge rectifier DC power supplies ordinarily found preceding PWM regulator systems. By the way, in some you'll find that bridging a capacitor across a suspected bad one, as we have suggested, will not remove all ripple voltage. If you run into this problem, it's an indication that the capacitor you're checking is leaky (has a high power factor), instead of being open.

How to Measure a Regulated Supply's Regulation

Power supply regulation is usually expressed as a percentage and can be calculated after making two voltage measurements. The first voltage measurement is made by connecting a voltmeter across the output terminals of the power supply, with a full load connected, as shown in Figure 3-7 (page 74). The second measurement is made by disconnecting the load resistor.

When you make the full-load voltage measurement, you'll have to adjust the load resistor to draw the power supply's rated full-load current. This value either can be calculated by using $I = E/R$, or measured. The maximum full-load current usually is given in the power supply operations manual, or you can find the value in many catalogs such as Heathkit's or Radio Shack's. The second measurement is made under no-load conditions. To do this, simply disconnect the load resistor and measure the output voltage of the power supply. Be careful that you *do not* make any changes of the power supply controls between measurements.

Now, all that is left to do is insert the two readings into the formula:

$$\% \text{ regulation} = \frac{\text{no-load voltage—full-load voltage}}{\text{full-load voltage}} \times 100$$

Incidentally, you can make the same test on power supplies with or without a regulation system. Also, as you can see by examining the formula, a low percentage of regulation is what you would want, because it indicates that there was very little change in the output voltage during the test. As an example, let's say that you measure a full-load voltage of 12 volts, and the no-load voltage is 15

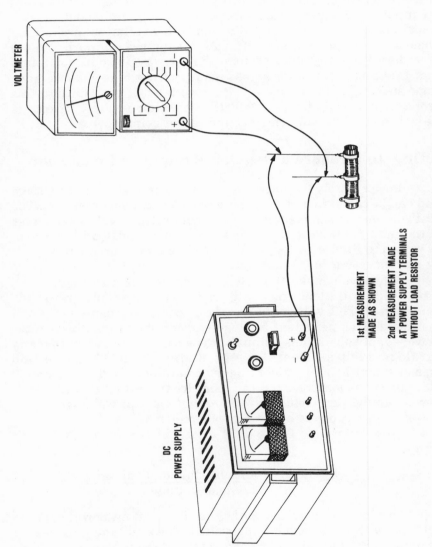

VOLTMETER

DC
POWER SUPPLY

1st MEASUREMENT
MADE AS SHOWN

2nd MEASUREMENT MADE
AT POWER SUPPLY TERMINALS
WITHOUT LOAD RESISTOR

Figure 3-7:   Voltage regulation measurement test setup.

volts. The percent of regulation would be (15-12)/(12) × 100, or 25%—in today's world, *very poor* regulation. Something like 0.05% voltage regulation is considered lab quality and a good supply regulation should be less than a 1% variation from no-load to full-load.

How to Test with a Wattmeter

Connecting a wattmeter into the AC power lines feeding a DC power supply, probably is the most useful single test you can perform—certainly, the simplest. By referring to the service data or another properly operating system, you can tell very quickly whether there is trouble in the DC power supply and, with just a little logical thinking, what the trouble is. A normal wattage reading, if the power suppy is disconnected from its load, tells you that the trouble is in the load. So you don't have to waste time checking the power supply or its regulating circuits.

A wattage reading quite a bit more than normal (with the load disconnected), shows that there is a short, or very high leakage, somewhere in the DC supply system. Ohmmeter checks usually will locate the faulty component in a very short time. Suppose, with the load disconnected your wattmeter reading is considerably lower than the normal reading. Your first thought is ... an open circuit (diode, resistor, choke or open filter capacitor, for example). A few quick voltage checks across these items, with a scope or voltmeter, just isn't good enough for checking regulated DC power supplies such as PWM supplies. In these, the waveshape (normally a rectangular wave) can be critical to proper operation. Therefore, it's important that you either compare the pattern you see on the scope to a properly operating supply or to the manufacturer's service notes.

How to Analyze Unusual Transistor
Circuit Systems

As shown in Figure 3-8, a defective transistor can produce puzzling symptoms during troubleshooting. We all know a transistor can develop an internal short or open. However, quite a few technicians don't know that it's possible to have different symptoms for different internal malfunctions. For example, if the transistor develops an internal open between the collector and base you'll find no output signal, as you would expect. This is shown in Figure 3-8 (A). But what about an internal short between the same two junctions? In this case, feeding a rectangular wave or sine-wave input to the base can very well result in a weak output signal at the collector. One way to know for sure if this is your problem is to check

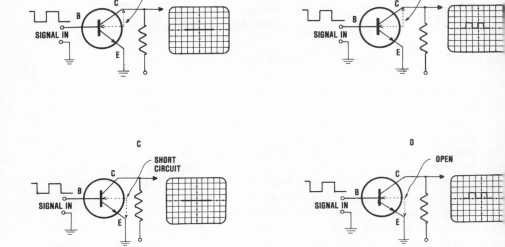

Figure 3-8:   Troubleshooting symptoms caused by internal open and shorts of transistor junctions, that you would see on a scope

the output with a scope. Compare the signal on the base to the signal on the collector. If both signals have the same phase (are not inverted) and the output peak-to-peak value is below normal, it's very probable that the transistor has an internal short between the base and collector, (see Figure 3-8(B).

It's also possible to get a small in-phase feedthrough if the base emitter junction is shorted (see Figure 3-8 (C). However, in this case, you should not find a signal on the collector output terminal. On the other hand, if the base emitter junction is open, more than likely you'll be able to see a definite feedthrough signal. It will probably be reduced in amplitude and be the same phase as the input signal on the base. But it should be easy to see on your scope (see Figure 3-8 (D).

Here's another trouble that you may have already run into. Have you ever connected a voltmeter across a transistor emitter resistor and the circuit started working ... then, when you disconnected the test leads, the circuit stopped operating? If you have, you already know that it can stop all production while you try to figure out what's happening. Here's what happens in this case. If the emitter resistor is *open* when you connect your voltmeter test leads across it, it completes the emitter circuit through the internal resistance of the voltmeter and, of course, the circuit will operate as

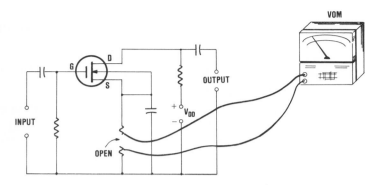

Figure 3-9:   How it's possible to start a non-operating N-channel MOSFET (depletion mode) circuit by connecting a voltmeter across an open source resistor. The voltmeter's internal resistance completes the open circuit.

long as you leave the leads connected. Figure 3-9 is a schematic example of what happens when you try to check a depletion mode **MOSFET** with an open source resistor, and, as you can see, removing the test leads will open the circuit again. This will, of course, bring circuit functions to an abrupt halt. This can happen in circuits using bi-polar transistors, **MOSFET's**, junction **FET's**, unijunction transistors, and other solid state devices, so it's a good thing to remember during troubleshooting of all similar components.

How to Test Periodic and Random Deviation (PARD)

The "DC output" of rectifier circuits such as shown in Figure 3-4 consists of *pulses* of current or voltage. As you know, the output has *ripple* and we must get rid of it. In today's equipment, there are two simple ways of doing this; 1) use the well-known pi-filter, and 2), use capacitors so big that they literally swamp out the ripple. As has been explained in a previous section, very large capacitors provide a path, with virtually zero AC impedance, to ground. For all practical purposes, the ripple flows through the filter capacitors to chassis or isolated ground.

You'll notice that, so far, the term *ripple* has been used to describe the small AC voltage found riding on top of the DC output voltage of the system under investigation. However, the current trend is to phase out the use of the term "ripple," as you may have noticed when reading various electronics publications. More and more, you'll probably find the acronym PARD being used by manufacturers when giving specs on their power supplies. The reason is that the measurement is made while holding certain external operational and environmental parameters constant. The exceptions are line and load transients, electromagnetic inter-ference, and drift.

The most important parameter that must be held constant during the measurement is the load impedance. Of course, this means that the temperature must also be held constant. PARD is measured under these constraints, with nominal line voltage and 50% load current. Practically speaking, the equipment is first brought to normal operating temperature by allowing sufficient warm-up time (then held constant) and, next, the periodic and random deviation of the output DC voltage, current, or power is checked for deviation from its average value with a scope or other instrument. When using a scope, measure the worst case peak-to-peak voltage. Then you can check your findings against the manufacturer's specs. However, in most cases, the manufacturer's line voltage is free of transients, will have very little (if any) drift, and EMI will be eliminated during measurements, as should your measurements be if they are to be meaningful. You also can use an rms responsive instrument to do the job. PARD may be specified as rms PARD or peak-to-peak PARD. The various signals that are included in the concept of PARD are shown in Figure 3-10.

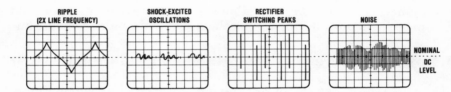

Figure 3-10:   The various signals that are included in the concept of PARD. It is assumed that all *external* operational and environmental parameters are held constant during the measurement.

Troubleshooting Power Supplies That Include
IC's in Their Design

As has been explained in the previous sections of this chapter, by far the most common rectifier circuit used in today's DC power supplies is the solid state full-wave bridge. However, these regulators are frequently constructed using **IC's**. The use of **IC's** in regulated DC supplies generally leads to a smaller and lighter overall package, which, in turn, drops the cost of the equipment (speaking strictly theoretically). For example, a tiny voltage regulator, such as a Signetics 5V regulator designed for local "on card" regulation, lends itself well to on-the-spot regulation. There are actually several types of these regulators manufactured by

Signetics, as well as other companies, that you may encounter during your troubleshoting (you'll find some are current regulators and some are voltage regulators). Figure 3-11 shows a typical application of a fixed 5V regulator (Courtesy Signetics, Menlo Park, California.)

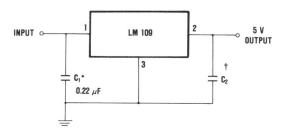

Figure 3-11: A small single silicon chip IC (TO-3 or TO-5 package) regulator, frequently used for local "on card" voltage regulation

The idea in using regulators such as these is to eliminate many of the noise and ground loop problems associated with single-point regulation. They generally employ internal current limiting, thermal shut-down, and safe-area compensation, which make the circuitry essentially blowout proof. You'll find heat sinks used in some instances and, as in all cases of this type, adequate heat sinking is essential for the protection of the device during troubleshooting or normal operation.

An example of a monolithic **IC** (Signetics 550) being used in a switching regulator is shown in Figure 3-12. The abbreviations on the **IC** represent the following parameters: V_{in} is unregulated input; V_{ref} is reference voltage; N.I. is noninverting input; V^- is negative terminal; COMP is frequency compensation; Inv is inverting input; CS is current sense; CL is current limit; and V_{out} is voltage output.

Although quite a bit has been said about switching regulators in previous sections, in this case the **IC** (the 550) is designed for the particular job of a precision voltage regulator. You'll remember that a switching regulator can produce excellent results and is used in TV receivers, etc. In fact, the efficiency of these systems may be in the

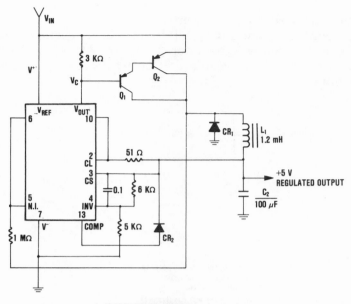

L₁ IS 50 TURNS OF #22 WIRE
WOUND ON FERROX CUBE.

Figure 3-12: The internal circuitry of the 550 IC permits ON and OFF strobing with diode transistor logic (DTL) and transistor transistor logic (TTL) inputs

80% to 90% range. This is because the switching transistor (also called *pass transistor*) is either hard ON, or in the OFF state. In practice, there is some loss, of course, because the transition between the two states is not instantaneous, but the loss is very small compared to a conventional DC power supply.

When you're troubleshooting this type of power supply, you'll find a significant difference between the switching regulators designed around **IC's** and those that have been constructed using discrete circuitry, in that the **IC** version is usually self-oscillatory. The discrete circuitry may use pulse-width modulation by the error signal to cause the switching transistor to alternate its conductive states (see Figure 3-1).

Another difference that you'll probably encounter is the operating frequency. The discrete circuit switching regulators normally operate in the several-thousands hertz region, while the monolithic versions can be operated at much higher frequencies (for example, 20 kHz to 100 kHz). In general, you'll find the higher frequencies permit a further reduction in component size, weight, and cost, but they also can cause troubles for the technician unless he takes certain precautions.

If replacement parts are required during repair, it must be remembered that ordinary power transistors and rectifier diodes cannot follow the high switching rates used in these power supplies. If you try them, not only will it degrade the performance, but you may, due to high power dissipation in the component, burn out the replacement part. Also, you'll notice that in Figure 3-12, the inductor L_1 calls for core material of Ferroxcube. The reason is that this material (you'll also find that Molybdenum-Permalloy toroidal cores are sometimes used) has a slow magnetic saturation rather than a fast one like ordinary silicon-steel laminated cores. You should avoid core materials with abrupt-saturation because it's possible there may be transients that can destroy other components. As we have said before, "When troubleshooting switching circuits, it's best to use exact replacement parts."

CHAPTER 4

Servicing Modern Solid State

Television Circuits

You'll find this chapter contains the latest TV color technology, including a comprehensive troubleshooting guide for common television troubles. It has just about everything you'll need to know about servicing, repairing, and adjusting color TV, including some of the latest solid state circuits and newest color picture tubes.

Today, color televisions using all-electronic TV tuners have become commonplace, and a great percentage of TV service jobs are for sets with this type tuner; output transformerless stages (OTL) and switching-type DC power supplies. There is much similarity between the OTL circuits used in TV receivers and the ones used in modern audio systems. In fact, you'll find that Chapters 1 and 3 outline troubleshooting procedures readily adaptable to servicing most contemporary TV circuits. Furthermore, as we all know, integrated circuits have become a part of every serviceman's daily job. Therefore, it's impossible to discuss troubleshooting anything electronic without including procedures for working with these devices.

The troubleshooting procedures for modern TV receivers are similar to what you have been using on the older sets. First, you turn the set on and analyze the picture on the screen (assuming there is a picture). As in the past, generally you can make a fair judgment of which section in the receiver is most likely to be at fault. However, there are some tricky symptoms that you have to watch for that will be explained in the following pages.

The main thing is to not jump off too quickly. It's very important to thoroughly analyze the picture on the screen *before* you start troubleshooting. It's easy to analyze incorrectly (as you can imagine, I'm speaking from experience) and end up spending half a day tracing signals in the wrong section of the set. One nice thing is that many solid state receivers now use circuit modules (making it less time-consuming when one gets lost) because it's faster to substitute a complete circuit.

Essentials of Troubleshooting Present-Day Solid State Television Receivers

Your very first step when troubleshooting any TV set is to turn the set on and, with the color control set at minimum, tune in a station. Check the set looking for raster, vertical problems, horizontal malfunction, etc.; in other words, a good monochrome picture reproduction. If you can get a good black and white picture on all channels, everything in the stages associated with the reproduction of the monochrome picture is functioning properly.

In troubleshooting the color processing in a TV set, all the controls must be checked for possible misadjustment first. *Don't even think of removing the back of the set until you have checked these controls!* You'll find that color trouble can appear as several symptoms; no color, wrong color, or loss of color sync. If you are using a color bar generator, check *its* adjustments and controls. For example, if the two pieces of equipment (TV and bar generator) are not on the same channel, you might work all day trying to get satisfactory readings and presentation.

Troubleshooting the color sections of a solid state TV depends on which type circuits are used (for example, transistor, **IC**, or a combination of both). In most late model sets, you'll find the color signals are processed by several **IC's**. In fact, if the trouble is a complete loss of color, a good starting point is to check the **IC's**. The easiest way to do this is by substitution. If you can't get an **IC** out (because of a soldered connection), use signal tracing methods. My suggestion would be to start by using your scope to check the output of the last **IC** used to feed the chroma detector. If you don't find a signal here, move back to the **IC** chroma input pin (this **IC** usually is called the *color processor*). Obviously, if there is a signal present on the input pin but not on the output pin, the **IC** can be considered to be bad. However, before you make your final decision to replace it, use your voltmeter and compare the voltages on the various pins against a schematic or a properly operating set of the same make and model. In some sets, you'll find the complete color sections contained in one or two plug-in modules. In this case, it's much simpler to substitute a known-to-be-good module, if at all possible.

Troubleshooting Monochrome Circuits

If you check a set and a perfect black and white picture appears, this clears everything in the set except the three basic color problems; no color at all, wrong color, and the loss of color sync. Furthermore, color problems can be narrowed down to just three separate circuits—the bandpass amplifier, the 3.58 MHz reference oscillator, and the color AFPC (automatic frequency and phase control) circuit. This means that once you have the black and white circuits operating perfectly, 90 percent of your job is done. Black and white or color television troubleshooting is primarily a diagnostic process and, as was mentioned before, the initial phase consists of evaluating the symptoms present on the screen of the picture tube and listening to the sound coming from the speaker.

The first thing to do when you start your troubleshooting procedure is to tune in a TV station and set the receiver color control for no color. If you get a good, clean, black and white picture, it means that the luminance signal is being processed correctly by all circuits, right up to, and including, the picture tube. If there are troubles, you service the monochrome section of a color set the same as any conventional black and white receiver. Just in case you are not familiar with basic TV troubleshooting, there is a list of the most common symptoms and their probable causes at the end of this chapter.

Special Testing and Servicing Factors

Modern TV circuits can develop some peculiar symptoms; loss of both syncs, pale video, and hum bars in the picture all at the same time or, just as strange, a vertical sweep that produces only half a picture on either the top or bottom of the screen. Tackling problems such as these takes a little special "know-how." For instance, if you run into multiple symptoms such as those just described, it's very possible that you have a DC power supply problem. This is particularly true in modern solid state sets because their supplies are very closely regulated. See Chapter 3 and *remember*, these power supplies are *critical* and should be one of the first things you check. Most of these TV's have shutdown circuits in their DC power lines and they are set to trigger at just a few volts rise. Because of this, you should check the set's schematic for the actual voltages used in a particular set. If these values are not correct, the shutdown circuit will go into instant action.

The other trouble, seeing raster on half the viewing sceen, also can be puzzling if you're not forewarned. As was brought up in the beginning of this chapter, many OTL circuits are almost identical to the ones described in Chapter 1. Generally, what has happened in the case of an OTL stage is that one of the transistors in its circuit has developed an open. The reason for the loss of half the raster is that the two transistors are biased to operate Class B and one produces the top half vertical sweep and the other, the bottom half. Figure 4-1 shows the two waveforms in a typical vertical OTL circuit that you should see on your scope, if you connect the scope across each of the transistor's emitter resistors.

Another odd problem you may encounter when servicing solid state vertical stages has to do with vertical linearity. For example, referring to Figure 4-1, you'll find a capacitor connected in the base-

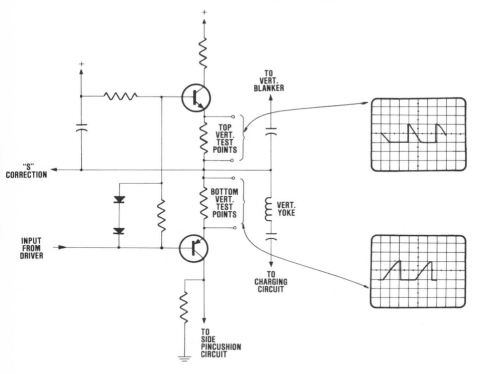

Figure 4-1:   Typical waveforms you should see on your scope when checking a vertical OTL. Similar OTL circuits are shown in Chapter 1.

emitter circuit of the top transistor. If this capacitor opens, it will cause linearity problems. The reason is that the capacitor provides positive feedback that slightly increases the gain of the top transistor. By the way, you'll find that feedback is used in all solid state vertical circuits. Troubles in these circuits can be caused by resistors changing value or by open by-pass capacitors, etc. The easiest way to troubleshoot these stages is to use a scope and check the waveforms for correcting voltage values and distortion. Simply follow the feedback loops and you should locate the trouble area very quickly.

Key Steps to Troubleshooting
Triac ON/OFF Switches

Several sets that do not use the on/off switch most of us are familiar with are on the market. Generally, you'll find that a triac has been used to replace it. This device can be described as a bidirectional switch (basically, two **SCR's** in parallel). The triac can be triggered into either forward or reverse conduction by a pulse applied to its gate electrode. The gate voltage normally used is very small and is provided in several ways. If you're working on a Quasar, you'll probably find it uses an optical coupler driven by the control **IC**. Others use an **IC** to switch the triac gate voltage.

Troubleshooting these on/off switches is easy. If the set stays on all the time, it's possible that the triac is shorted. Figure 4-2 shows how to test a triac using an ohmmeter. First, connect the ohmmeter leads to anodes 1 and 2. You should read a very high resistance. Next, make a make-and-break connection with a jumper lead between anode 2 and the gate. You should see the triac resistance drop to a low value and stay there, even when you remove the jumper lead. This completes the check for one-half the triac. To test the other half, simply switch the ohmmeter leads and again short between the gate and anode (number 2). This half should show a low resistance reading in the opposite direction.

There are some triacs that have a built-in diac gating element. In case you run into one of these, it's necessary to use a larger voltage than the average VOM puts out. You'll need about 22 volts DC before you can trigger them. Switch your VOM range switch to the 100 mA range (in this case) and insert a 220-ohm resistor in series with the meter and DC supply. Other than this change, the test procedure is the same. By the way, in reality, polarity is not important in this check (nor the first one), just as long as you switch leads in the second step.

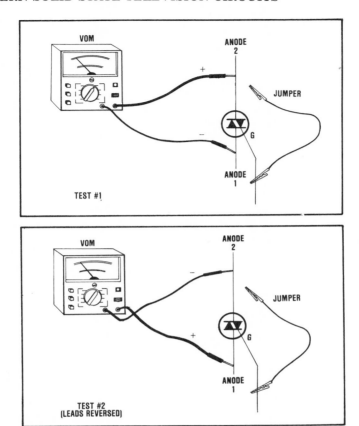

Figure 4-2:   Test setup for checking a triac using an ohmmeter

If the set will not turn on at all, check the gate voltage. There may be a problem in the control circuit. The other possibility is that you have an open triac. To check for this, use the same procedure as explained for checking a shorted triac. In some TV receivers, you'll find that a microprocessor is used to control on/off, plus other functions. The easiest way to begin, in this case, is to check if you have the correct DC voltages at the test points. Make sure that they are *exactly* what the schematic calls for, because they are extremely critical.

Servicing the New Solid State TV Front Ends

The function of the television receiver front end is the same in all TV sets, old or new. However, the physical appearance and electronic circuits have changed. All-electronic TV tuners, using solid state bandswitching and tuning, are being found in more and more receivers. You might say that the heart of these TV front ends is a varactor-tuned oscillator. Typically, the local oscillator coil is tapped. The most important point in troubleshooting is—the entire coil is normally used to tune over the low VHF channels and the top half is used for the high VHF channels (the total range of frequencies is 54 MHz to 216 MHz). The basic idea of troubleshooting a varactor-tuned circuit is shown in Figure 4-3.

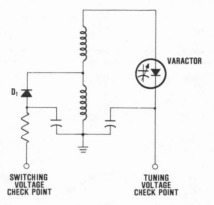

Figure 4-3:   Control voltage check points for a typical varactor local oscillator

In the low-band mode, the switching diode (D₁) is nonconducting because a control voltage is being applied to its anode (in this case, the control voltage must be negative to shut the diode off). As you can see, testing is simple. Read the set's schematic and determine what the voltage should be in the low VHF mode. Next, measure it at the point shown. You'll probably find the voltage should be 20 or 30 volts. If it is incorrect, refer to the schematic to find out where it comes from and check that stage. Chances are that it's coming from a separate regulator.

To check the switching diode for high-band operation, measure the voltage in the same way. But, in this case you should find a positive voltage at the check point shown in Figure 4-3. Remember, this is only an example. You may find different polarities on the switching diodes in different sets. As always, the first step should be to check the schematic. Figure 4-4 is an example of a varactor-tuned oscillator stage. Don't overlook the possibility of a shorted switching diode. If this is true, it may affect the high-band VHF stations but not the low-band ones. To check the diode, lift one end and measure the front-to-back resistance with an ohmmeter.

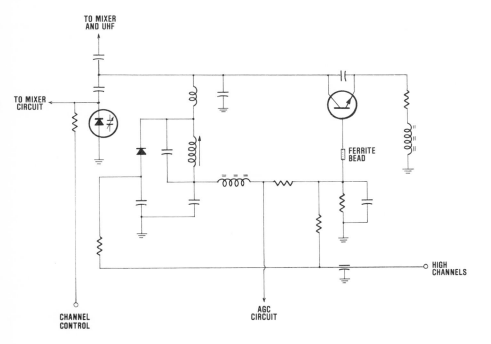

Figure 4-4:   Simplified schematic of a varactor-tuned solid state TV receiver local oscillator

The best troubleshooting guide is the manufacturer's service notes. These usually include the voltage range for specific frequency bands and some even give you the correct voltage reading for each channel.

As was explained in Chapter 3, most DC voltages in modern electronic equipment (such as hi-fi and TV sets) are tightly regulated, and these TV circuits are no exception. If the varactor is off, check the regulator. It's also possible that there is a problem between the regulator module and the DC voltage supply.

If the set is controlled by a microprocessor, you'll find circuits such as phase-locked loops (PLL) that note any phase difference between signals, then convert it into a correction voltage that causes the phase of the output voltage to change so that it tracks the reference. When troubleshooting these receivers, all DC voltages are *very critical*. If the tuning voltages are not correct, the set will be tuned off-station. Most manufacturers use varactors to tune several stages in the front end, with or without use of a micrprocessor. Some examples are antenna tuning and local oscillator tuning. Figure 4-5 is a simplified schematic diagram of a typical rf pre-amplifier and its feed lines.

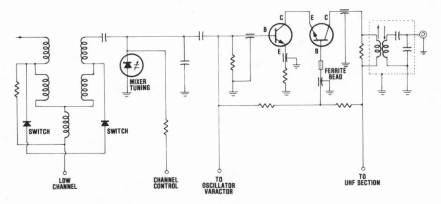

Figure 4-5:   Schematic diagram of a solid state varactor-tuned pre-amplifier

Key Steps to Troubleshooting Solid State
Bandpass Amplifiers

The typical solid state chroma bandpass amplifier is designed to separate the chrominance signal from the composite color signal and feed it to the demodulators. The input signal usually comes from the first video amplifier as a composite signal that includes the chrominance, luminance, burst, sync, and blanking signals. The circuits may be constructed using transistors and/or integrated circuits. In the circuit shown in Figure 4-6, the composite video signal from the first video amplifier is fed to two chroma IF amplifiers, where it is amplified before being coupled to the bandpass amplifier by the bandpass transformer. The bandpass transformer primary is a tuned circuit that ensures the proper bandpass for the stage.

In some instances, you'll find a resistor connected across the resonant circuit. The purpose of the resistor is to broaden the resonant response, thus enabling the load impedance to accommodate the range of frequencies desired. What you want is to make the circuit broadly resonant around the 3.58 MHz region so it can pass the desired chrominance sidebands.

When you are troubleshooting a bandpass amplifier (see Figure 4-6), you should measure the voltages, check for proper input signal, and test the transistors (or **IC's**, if used). Because the bandpass amplifiers have gating pulses applied to them, you should examine these pulses with a scope. At the same time, check the service notes for the set you're working on. They will list the correct

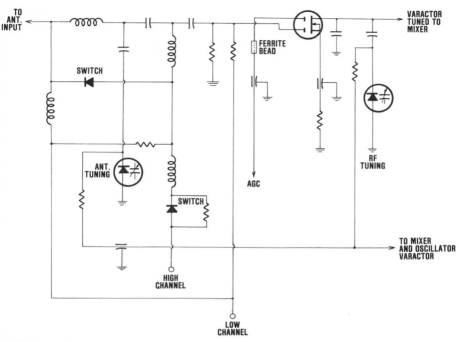

Figure 4-6:   Solid state chroma bandpass amplifier

peak-to-peak values of the pulses you should see at various points in the circuit.

A complete loss of color can be due to several causes. For example, a "dead" 3.58 MHz oscillator, or a trouble in the color killer are two common causes. To localize the bad circuit, first feed a color bar generator signal into the TV set and look at the bandpass amplifier's output on the scope (in some sets, you'll find much of the color processing circuitry, including the bandpass amplifiers, is contained in an **IC**). In most cases, you can't find this signal at the color control, but you can always find it at the input to the demodulators. At any rate, if you see a normal "comb" pattern with the correct amplitude on the scope, the bandpass amplifier stage and color killer are working correctly. You can also check the oscillator with your scope. Simply look for the correct peak-to-peak signal.

I strongly suggest that you don't try color alignment except as a *last resort*, and only after you have checked and checked again. *Remember,* alignment has to be really bad before it will cause a complete loss of color. Furthermore, to the best of my knowledge, it will never cause a sudden loss of color. If you see good clean bars on the scope, the set probably doesn't need color alignment. Incidentally, to assure a clear pattern if you're using an older scope, it's usually best to use a crystal-detector probe when you're signal tracing in the bandpass amplifier stages. If your scope is wideband, simply use your low-capacitance probe.

How to Check the Color-Sync Circuits

As it is for other electronics circuitry, the scope is a handy instrument for checking signals (or the absence of) at the input and output of the sync stage. This is particularly true when checking a color sync circuit using an **IC**. In fact, I think you'll probably find that the best approach to servicing modern solid state color sync circuits is to use signal-in, signal-out methods and a good scope.

If you're working on transistorized circuits, start off by touching the scope probe to the various components. For example, begin at the sync separator output, then check the burst amplifier transistor collector, and on down the line to the 3.58 MHz oscillator and amplifier, ending up at the input to the color demodulators. A color-sync circuit using an **IC** will not have as many check points, but who's complaining? This makes our servicing jobs easier.

Be careful with the 3.58 MHz signal because it is critical. In modern TV sets, a properly operating oscillator/AFPC circuit holds the frequency so steady that, using ordinary test gear, there is no detectable phase shift. Why is this so important? Because a phase shift of *one-sixteenth of one cycle* causes a change in color; for instance, strangely hued people.

In these sets, you'll find the oscillator frequency is controlled by a DC voltage. If you need to adjust it, place a jumper wire on the AFPC circuit and remove its control. Then adjust the reactance coil, etc. Refer to the set's service data for complete instructions. *Follow these instructions to the letter* until you see the color lock in momentarily. Next, remove the jumper lead off the AFPC circuit and you should see the color lock in. If it doesn't, there is a trouble in the AFPC circuit. Another circuit that is easy to overlook and can cause you to have color-sync problems is the horizontal oscillator AFC. If you're having troubles with lock-in, it's a good idea to check the adjustment of this stage.

Color Killer Test Procedures

If you have a complete loss of color, it can be due to a fault in the color killer circuit. Of course, it could also be a trouble in the 3.58 MHz oscillator circuit, as well. Therefore, the first step is to determine which circuit is causing the problem. To isolate the bad circuit, feed a color bar signal at the input to the demodulators and, if it is the right voltage, the color killer circuit is performing correctly. It also means the bandpass amplifier is working.

If you don't see a normal "comb" pattern on the scope, check the bias on the bandpass amplifier that is controlled by the killer circuit (for example, the second bandpass amplifier). You should see this voltage drop quite a bit when you feed a color signal into the set. Some servicemen like to over-ride the killer bias to make sure the color signal can get through. If you try this, and it works, it's almost certain that the trouble is in the color killer circuits.

At this point, **stop!** Ask yourself, "Did I check the controls? Could it simply be a misadjustment?" As soon as you have assured yourself it isn't an adjustment problem, check out the killer bias circuitry. Check for bad diodes, transistors, **IC's**, etc. You'll find that most of today's color receivers employ **IC's** in the chroma circuits. In these sets, the color killer circuits usually are included in an **IC**. The tests are quite simple. In most, the only instrument you'll need is a DC voltmeter. A simplified solid state color killer circuit is shown in Figure 4-7 and includes some typical voltage amplitudes.

Practical Color Demodulation Servicing

In most late-model color TV receivers, you'll find that **IC's** are used in the signal processing circuits. Figure 4-8 shows a simplified color demodulator circuit using an **IC** and some typical voltage

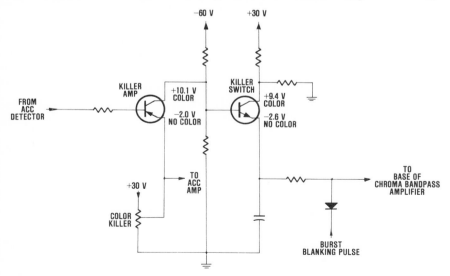

Figure 4-7:   Typical voltage amplitudes found when servicing solid state color killer circuits

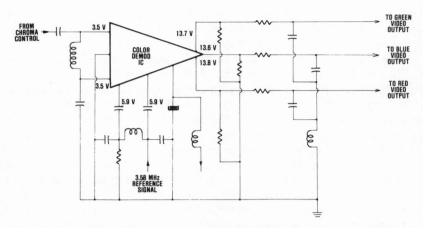

Figure 4-8: A color demodulator stage using an IC, showing typical voltage amplitudes found in the circuit

amplitudes. The DC voltages shown on the service schematic, as well as this simplified one, are for no-signal conditions (meaning, no color).

If you're getting weak tints (such as color bars that, lacking red, appear bluish and greenish), it usually means the trouble is not in the demodulator circuits. If you have these symptoms, check the oscillator. In many sets, a "dead" oscillator will cause these exact conditions. Like most **IC** circuits, the active components for the demodulator stage are contained within the **IC**. Therefore, troubleshooting a color demodulator circuit using an **IC** is limited to checking voltages on the **IC** pins and using your scope to observe input and output waveforms.

Troubleshooting the Matrix Section

Matrix means different things to different servicemen. For example, a coding system to computer people, an impression from which a large number of phonograph records can be duplicated or, in a color TV circuit, the section that combines the color difference signal at the output of the chrominance demodulators, which is the one we are interested in. The idea is to combine the signals in the proper proportions so that the correct color signals are produced. In the end, the color signals are applied to the picture tube where they reproduce the hues of an image in terms of the three primary colors— red, green, and blue. The matrix circuits shown in Figures 4-9, 4-10, and 4-11 are from a solid state color receiver and show some typical voltage amplitudes for various check points.

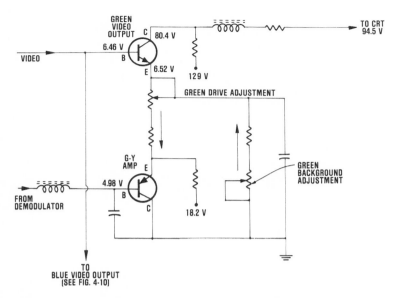

Figure 4-9: Solid state color receiver green matrix circuit showing typical voltages at various test points. All voltages are measured under no-signal conditions.

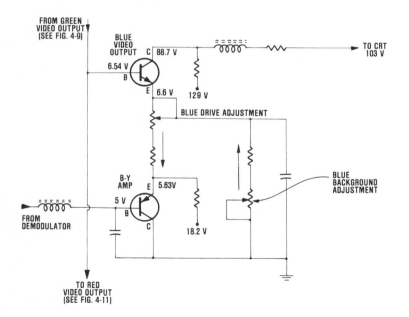

Figure 4-10: Solid state color receiver blue matrix circuit showing typical voltages at various test points. All voltages are measured under no-signal conditions.

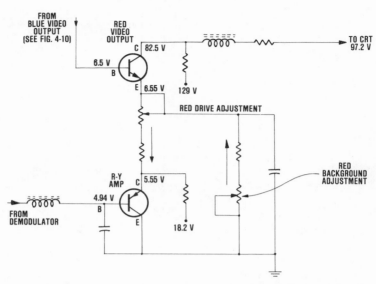

Figure 4-11:   Solid state color receiver red matrix circuit showing typical voltages at various test points. All voltages are measured under no-signal conditions.

You'll find that quite a few solid state color receivers use separate demodulators for each color channel. For one reason, it simplifies the matrix circuit. For another, it works well when there are **IC's** being used. When troubleshooting a solid state matrix system such as the ones shown in Figures 4-9, 4-10 or 4-11, one of the best ways to start is to scope the inputs to the three amplifiers labeled G-Y amp, B-Y amp, and R-Y amp. You should see three signals with very nearly the same pattern. The peak-to-peak voltages should be near the same amplitude (5 or 6 volts peak-to-peak), except for the R-Y amp, which probably will read slightly over a volt (peak-to-peak).

If we could just say "Hail and farewell" to all adjustments, many of us would be much happier. But the world doesn't seem to work that way. In spite of all modern engineering efforts, a few adjustments seem to make it past the drawing board. To make matters worse, a customer with a screwdriver inside a color TV— working on the adjustments—can be more dangerous than a small child with a hammer. Why emphasize this? Because it's *very important* that you *check the adjustments*.

Referring to Figures 4-9, 4-10 and 4-11, you'll see the emitter bias for each of the video output stages is determined by an adjustment. In fact, you'll see six adjustments in all. But, you'll find they are easy to set up. Simply adjust them to produce a good *black and white* picture at all brightness levels. If you can't get a good black and white picture, check the voltages (or scope the circuits). These tests usually are quite simple; in most, your voltmeter is about the only instrument you'll need.

How to Service the Latest Color Picture Tubes

Usually, with the help of a CRT tester/rejuvenator and performance data to substantiate his observations, a serviceman can visually diagnose unsatisfactory picture tube performance. Like all tube testers, a CRT tester/rejuvenator isn't necessarily the final word. For example, to know if rejuvenation has been successful, it is necessary to measure the quality of the beam current after you've completed the rejuvenation procedure. Unfortunately, you'll find that many of these instruments indicate the grid current, which does not let you know what the emission quality of the cathode is in the most important location—directly under the grid aperture. The only way to make this test is to make a true beam current-emission test. Incidentally, this is why you'll sometimes find a CRT that checks out good on a CRT tester and yet fails to perform properly in the set.

Another trouble you may run into is that when you rejuvenate a three-gun color CRT, restoring one gun may require one (or both) other gun to be rejuvenated. Because this can happen, it's better to use an instrument that includes clean and rejuvenate modes for each gun ... *if you are using a CRT tester/rejuvenator.* This last statement would lead us to believe there is another way to check a picture tube and try to increase its useful life, and there is. You can always try a tube brightener with no risk (which isn't so with all CRT rejuvenators). They are available at your nearest electronics supply store (Radio Shack, etc.), and much less expensive than tester/rejuvenators. Placing a picture tube brightener in the set will quickly tell you, in most cases, if the picture tube is the problem.

Testing Modern TV Power Supplies

Generally speaking, you'll find servicing modern TV power supplies will require a lot more use of the scope than the older type power supplies. For instance, you must use a scope to trace the control pulses through the pulse width modulated (PWM) DC power supply that was covered in Chapter 3. It is absolutely essential that these pulses go to the right place, *with the correct waveform!* As mentioned before, if the entire PWM supply appears to be nonoperative, but the unregulated DC voltage from the rectifiers is okay, start your troubleshooting in the *horizontal oscillator* stages.

You'll find that most high voltage shutdown systems used in solid state sets are designed using an **SCR**. The **SCR** is connected across the drive circuit of the switching transistor, whose gate senses the emitter voltage of the transistor. Let's assume something happens that causes the transistor to conduct too much current. What will happen is that the emitter voltage will rise and this will cause the **SCR** to start operating and short the circuit. In turn, the switch is caused to be nonoperative and will stay that way until the **SCR** is turned off (remember, an **SCR** stays on—once it's fired—until its anode voltage is cut off). You'll have to turn off the TV receiver to reset the high voltage shutdown circuit; i.e., the **SCR**.

Although the modern TV DC supplies are tricky, they are not difficult to troubleshoot if you know what they are supposed to do and how they do it (see Common Television Symptoms and Probable Causes at end of this chapter, and Chapter 3 for additional hints and troubleshooting procedures). One of the main things to remember is that, since the operation of transistors and other solid state devices can be seriously affected by small voltage changes, it is extremely important that the voltages be well regulated and ripple free. When in doubt, check with your scope. Incidentally, in TV sets with flyback-derived, low voltage power supplies, ripple will be at the horizontal frequency.

Common Television Symptoms and Probable Causes

SYMPTOM	PROBABLE CAUSE
No Picture—Sound Normal	
1. Bright horizontal bar showing on picture tube	Loss of vertical sweep. Check vertical deflection coils.
2. No picture, have raster	Trouble in stages after sound pickof
3. No picture or raster	1. Check high voltage power supply
	2. Check bias on grid of picture tub
	3. Trouble in flyback sweep system
No Picture or Sound—Have Raster	1. Check antenna connection and trans mission line
	2. Bad solid state component or part i *both* sound and picture sections of pi ture
	3. Trouble in the tuner (rf mixer-osci lator) or defect in any stage from rf t sound take-off
No Picture, Raster, or Sound	1. Check AC outlet and line cord
	2. Check on/off switch
	3. Check interlock on chassis
	4. Trouble in low voltage power suppl or low voltage system
	5. Bad solid state component or part i *both* tuner and high voltage syste

SYMPTOM	PROBABLE CAUSE
Sound and Picture Missing for Some Channels, Not for Others	Trouble in tuner or improper tracking
Weak but Good Quality Sound and Picture	1. Antenna wrong type or bad location 2. Improperly oriented antenna 3. Trouble in antenna system 4. Decreased voltage from low voltage power supply 5. Change in tuner tracking or IF alignment (may affect quality) 6. Defective solid state component or part in tuner or in *both* video and sound sections
Poor Quality and Weak Picture and Sound	1. Low line voltage 2. Check AGC system 3. Change in fine tuning control 4. Change in local oscillator frequency 5. Trouble in tracking and alignment of tuner and IF stages 6. Bad solid state component or part in *both* audio and video amplifiers
Weak or Absent Sound—Pronounced Diagonal Bar Interference on all Stations	1. Oscillating audio IF amplifier or an oscillating state prior to sound takeoff 2. Oscillating rf amplifier stage in tuner
Heavy Slanting Streaks Across Screen	
Sound normal	1. Horizontal sweep out of synchronization 2. Check receiver controls and/or horizontal sweep circuit
Rolling up or down	1. Check vertical or horizontal controls 2. Trouble in sync separator, clipper or sync amplifier circuits 3. Trouble in both sweep systems 4. May be sync clipping in video stages
Sound Poor or Missing— Picture Normal	
Hum from speaker	Could be the loudspeaker
Weak sound	1. Check fine tuning control 2. Check components in sound section 3. Could be a change of some component value in local oscillator causing drfit
Sound distorted	1. Check speaker cone and voice coil 2. Check bias in audio section 3. Check interstage coupling capacitor 4. Check solid state components 5. Shorted capacitor

SYMPTOM	PROBABLE CAUSE
4. Buzz and noise from speaker	1. Check high voltage arcing
	2. Check contacts in low voltage supply feed lines
	3. Trouble in tuner and IF stages
	4. May be alignment is off in tuner and/or IF stages
	5. Open large capacitor filter across ratio detector output
	6. Video carrier containing some FM
	7. Trouble in limiter
	8. Loose laminations, loose windings, or arcing within vertical output transformer

Incorrect Picture Size

1. Picture straight, but off center	1. Misadjusted centering controls
	2. Defective deflection
2. Insufficient picture width	1. Check width control
	2. Weak solid state component in horizontal amplifier circuits
	3. Trouble in deflection coil
	4. Defective parts in horizontal sweep system
3. Insufficient picture height	1. Check height control
	2. Defective deflection
	3. Defective vertical output transformer
	4. Weak solid state component vertical amplifier circuits or incorrect value resistors or capacitors
4. Excessive picture size in both vertical and horizontal directions	1. Both size controls misadjusted or defective
	2. "Blooming" caused by decrease in high voltage to picture tube
5. Picture too small	1. Misadjusted size controls
	2. Low line voltage or overloaded power mains
	3. Defective yoke
	4. Defects or degeneration in both sweep sections
	5. Decreased potentials from low voltage power supply
	6. Excessive high voltage
6. Keystoned picture with reduced width	1. Shorted capacitor across one-half horizontal deflection coil
	2. One-half vertical deflection coil short circuited
7. Keystoned picture with reduced height	1. One-half vertical deflection coil short circuited
	2. Shorted resistor across one-half vertical deflection coil

SYMPTOM	PROBABLE CAUSE

Interference on Picture—
Sound Normal

1. Slanted, tweedy lines across picture
   1. Interference from CB station or other electromagnetic interference
   2. Oscillating stage in receiver
2. Lacy (moire) effect in picture
   1. Could be loss of interlace
   2. Check for defective component in integrator circuit
3. Ragged bar across picture, "snow" in picture
   1. Weak station signal (fringe area)
   2. Poor antenna installation or open circuit in antenna system
   3. Improperly adjusted AGC system
   4. Defective AGC system
   5. Poor signal-to-noise ratio in receiver
   6. Defective solid state component in any stage from tuner to picture tube
   7. Improperly tracked tuner or mis-aligned video IF stages
4. Intermittent streaks across picture
   1. Ignition interference, poor signal-to-noise ratio in receiver
   2. Interference from nearby motors or other electrical devices
5. Intermittent streaks across picture when set is tapped
   1. Check connections at antenna terminal posts
   2. Poor socket contacts
   3. Poorly soldered part in chassis wiring
      Check for misadjusted 4.5 MHz trap in video amplifier stage
6. Thin, closely spaced vertical lines
7. Black vertical bars (one or two) near left of picture or raster
   1. Check for transient oscillations in horizontal output amplifier
   2. Check horizontal output section
8. White vertical bars (one or two) near left of picture or raster
   1. Trouble in damping circuit
   2. Trouble in voltage boost system
   3. Trouble in horizontal sweep output system
9. Vertical black bar down center of picture—picture in two sections, back-to-back
   1. Check phase in horizontal lock sync system
   2. Trouble with a solid state component in horizontal lock system
10. White vertical bar near center of picture
    1. Check drive control (left of picture stretched)
    2. Check for defective part in horizontal amplifier circuits
    3. Check components in damper of voltage boost system
       Check for interference above horizontal sweep frequency
11. Vertical bars on screen
       Check for interference below horizontal sweep frequency
12. Horizontal bars on screen
13. Horizontal bar at top of picture
    1. Check picture linearity
    2. Check damping resistor across vertical coil section

SYMPTOM	PROBABLE CAUSE
14. Audio bars on screen	1. Co-channel interference 2. Trouble with sound interference in picture 3. Misadjusted sound traps
15. Interruption of vertical detail by horizontal line structure	Check for interaction between sweep circuits
16. Framing vertical and horizontal bar pattern moving across screen	1. Check for picture carrier interference from upper adjacent channel 2. Check upper adjacent channel trap adjustment 3. Check antenna orientation and/or front-to-back antenna ratio
17. Repeat lines at right of white-to-black or black-to-white objects	1. "Echo effect" due to high frequenc transients in video amplifier stage 2. Overpeaked video amplifier 3. Incorrect value peaking coils 4. Trouble in solid state components (component part values in video amplifie
18. Retrace lines showing in picture	1. Check for weak signal strength 2. Trouble in picture tube 3. Improper blanking
19. Double image picture (ghost)	Interference from reflected signal
Poor Picture Quality	
1. Excessively dark picture	1. Trouble in AGC system 2. Check DC restorer circuit 3. Check contrast control
2. Dark picture with trailing smears	1. Check low frequency response in video amplifiers 2. Check tuner tracking 3. Could be video IF alignment 4. Check decoupler circuit in video amplifier stages
3. Negative picture	1. Trouble in picture tube or signal overload 2. Check AGC circuit 3. Check oscillator 4. Check peaking coil
4. Faded picture with normal contrast	Check emission of picture tube (depending on contrast setting)
5. Indistinct picture	1. Possible reflections on transmissi line 2. Trouble in picture tube 3. Check high voltage on second ano of picture tube 4. Could be improper alignment 5. Check peaking coils 6. Trouble in video amplifier stages 7. Check focus adjustment or focus electrode voltage

SYMPTOM	PROBABLE CAUSE
Only a portion of picture area in focus	Check voltage in focus electrode circuit
Intermittent picture shift to left and right	1. Trouble in horizontal centering control 2. Check horizontal lock 3. Intermittent voltage change in vertical system
Intermittent picture shift in vertical plane	1. Check vertical centering control 2. Intermittent voltage change in vertical system 3. Check vertical oscillator or output system
Slight weave at top of picture	1. Check vertical centering control 2. Check for proper sync amplitude 3. Could be sync slipping in video amplifiers
Picture weaving at both sides (plus excessive contrast)	1. Check AGC 2. Excessive signal strength at picture tube grid 3. Check horizontal lock system
Picture weaving at sides, contrast normal	1. Interference present in video signal 2. Ripple (AC) in horizontal sweep 3. Misaligned video IF stages causing sync amplitude decrease 4. Check low frequency response in video amplifier prior to sync takeoff
Picture weave on strong stations (or when fine tuning set for maximum picture signal)	1. Check contrast control 2. Trouble in AGC system 3. Check bias in tuner stages 4. Check bias in video IF stages 5. Check leveling in sync separators feeding synchroguide or phase detector
Picture weaving, poor contrast, snowy picture	1. Could be weak signal 2. Defective solid state component or part 3. Trouble in a circuit prior to sync takeoff 4. Misaligned tuner or IF stages
Unstable vertical synchronization	1. Check vertical hold control 2. Check setting of AGC 3. Trouble in AGC system (horizontal sync will be less affected) 4. Trouble with a solid state component or part in vertical oscillator and control 5. Trouble in sync separator to vertical input circuit
Unstable horizontal synchronization	1. Check horizontal hold control 2. Check horizontal lock system 3. Defective solid state component or part in horizontal oscillator and control 4. Trouble in horizontal system 5. Check sync pulse in stages prior to AFC system

SYMPTOM	PROBABLE CAUSE
16. Both vertical and horizontal sweep unstable	1. Check AGC system
	2. Check *both* vertical and horizontal sweep systems
	3. Could be poor low frequency response
	4. Check sync in video stages
	5. Check voltage amplitude to sync separator or sweep stages
	6. Defective diode, transistor, IC, or part in sync separator or amplifier stages
	7. Trouble with DC restorer diode or circuit from which signal is taken for sync separator stages
17. Intermittent picture and sound	Check components and connections from tuner to picture tube grid
18. Intermittent picture-sound normal	1. Check picture tube socket
	2. Trouble in video amplifier or picture tube
	3. Look for loose connections and bad components in IF stages *after* sound takeoff
	4. Look for loose connections and bad components in detector or video amplifiers
19. Intermittent picture, raster, and sound	1. Check for shorts at voltage source
	2. Check low voltage power supply system
20. Intermittent sound-picture normal	1. Check for arcing in high voltage power supply
	2. Check from sound takeoff to speaker
	3. Check high voltage solid state diode rectifier
21. Picture folds over at left	1. Check retrace and blanking intervals in horizontal sweep
	2. Check damping circuit
22. Picture folds over at right	Check low frequency response in sweep amplifier
23. Picture folds over at bottom or top	1. Look for a leaky coupling capacitor in vertical output
	2. Check solid state components and parts in vertical sweep system
24. Excessive fold over	Could be a loss of harmonic components in the horizontal sweep waveform
25. Picture stretches out at left but fills mask fully	Check drive and width controls
26. Picture information crowded at left	Trouble in horizontal system from discharge to deflection coils
27. Picture information crowded at right	1. Check horizontal linearity
	2. Trouble in horizontal system from discharge to deflection coils
	3. Could be a defective drive control (also causes stretching at left)
28. Picture information crowded at top	1. Check vertical linearity
	2. Trouble in vertical sweep system
29. Picture information crowded at bottom	Trouble in vertical sweep system

CHAPTER 5

Differential and Operational Amplifier

Servicing Guide

Today, the availability of inexpensive operational amplifiers (**OP AMP's**) has made them a practical replacement for *any low-frequency amplifier*. For instance, you can purchase a whole handful of them for about $2.00 (a "whole handful" being about ten to twelve 741's). By no means are these low-cost **OP AMP's** restricted to linear amplifiers. In fact, you'll find they are used in an incredible variety of DC and audio uses. Just a few examples are active filter circuits, precision rectifiers, integrators, ramp generators, electronic music circuits, and many more.

By now, everyone is probably more or less familiar with the 741 and most of its improved offspring such as the 5558 (Signetics), MC 1458 (Motorola), 4136 (Raytheon), and the improved LM 318 (Advanced Micro Devices, and National). In general, you'll find that most of these **OP AMP's** are available at reasonable cost from electronics dealers. They also are the type of **OP AMP's** you may encounter during troubleshooting electronics equipment manufactured in the past few years; i.e., the last half of the 1970's.

Many troubleshooting jobs deal with systems that are designed to amplify slowly changing signals or even DC levels. In Chapter 2, Figure 2-14 shows an **IC** operational amplifier that will work very well with this kind of input. However, you'll also find single and cascaded *differential amplifiers* used for jobs like these, especially in oscilloscope, electronic meter, and recording instrument amplifiers. The differential amplifier is also used as the

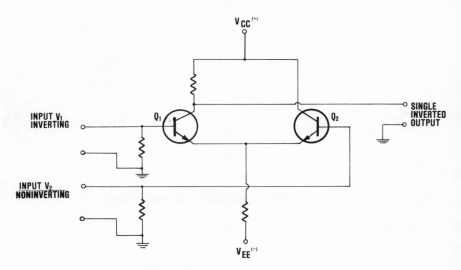

Figure 5-1:   Basic differential amplifier

input circuit of an operational amplifier. Figure 5-1 is the schematic of a basic solid state differential amplifier.

Differential Amplifier Performance Tests

One important characteristic of a differential amplifier, for the troubleshooter, is that an output signal should not be present except when there is a *big difference in signals at the input*. This performance characteristic is what makes the amplifier so useful because, when the signals are common to both inputs (generally referred to as *common-mode signals*), they are greatly reduced in amplitude or totally eliminated. Of course, this will cancel out such things as AC power line hum and temperature changes, which is why they are one of the most popular designs in use today.

The main point is that when you're checking a transistor circuit based on the simplified circuit shown in Figure 5-1, you should find the emitter circuit current at any point to be equal to any other point (assuming the collector voltages for Q_1 and Q_2 in Figure 5-1 are equal). In some circuits, you'll find a resistor in Q_1 and Q_2's collector leads. In this case, the voltage measurements across both the resistors should be very near the same. These measurements must be made with either no signal in, or when both input signals are *exactly equal*. Under these conditions, the output voltage should be about zero.

Working with Operational Amplifiers

If more gain is needed than can be obtained from a single differential amplifier, the manufacturer cascades several more stages to build what is called an **OP AMP**. This provides both common-mode rejection and high gain. A schematic symbol of an **OP AMP** and a simplified block diagram are shown in Figure 5-2.

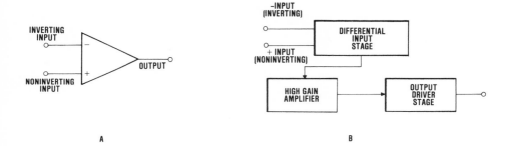

A   B

Figure 5-2:   OP AMP schematic symbol (A), and simplified block diagram of the interior of an OP AMP IC (B)

By referring to Figure 5-2, you'll notice there are two inputs to the differential input stage of the **OP AMP**. When reading a schematic, you'll normally find one input marked with a minus sign. This input is called the *inverting input*. The other input is usually marked with a plus sign and is called the *non-inverting input.* Although this was explained in Chapter 2, it's worth repeating because it brings up the subject of a split power supply. The schematic symbols of (+) and (−) on the signal input circuit have nothing to do with the power supply input. You may find the voltage supply pins marked V+ and V−, while other manufacturers may use V_{EE} and V_{CC} to represent negative and positive respectively. In other words, you must look at the schematic carefully before applying the amplifier power source.

In the so-called good old days, we usually thought that any power supply terminal marked negative was an earth ground point. This is still true when working with many circuits, but not for all **OP AMP's.** There is no ground connection for some **OP AMP's** and you'll find many use split supplies; for example, one +15 V line and one −15 V line. Furthermore, the input signal level must be restricted to working with significantly less than the full amplitude of either of the two supply voltages. To illustrate, in one case (a 741 **OP AMP**), the two supply voltages are +15V and −15V. The input signal level is limited to ±12 volts. You'll find this specification called the *common mode range*.

If the circuit you're working on is using both inputs, it is working in the *differential mode*. But if you find that one of the inputs is grounded (or tied to some other reference), it's said to be working in the *single-ended mode*. Next, when you are checking the schematic of a piece of gear using **OP AMP's**, look at the feedback lines. If you find that the feedback is from the output to the positive input, it's probably a digital operation. On the other hand, if the feedback is from the output to the negative input, the circuit is set up for a linear operation. The gain of the circuit depends on the ratio of R_3 to R_1, as indicated (see Figure 5-3).

Glossary of OP AMP Terms
Needed for Troubleshooting

Input Offset Voltage—This is the slight voltage difference (a millivolt or so) between the **OP AMP** inputs. Also, see Chapter 2, "Interpreting Current Data Sheets."

Input Offset Current—This refers to the difference in the two currents (measured at the two input terminals). Also, see Chapter 2, "Interpreting Current Data Sheets."

Input Voltage Range—Watch this parameter! It is the range of voltage you can safely apply to the input without damaging the device.

Output Voltage Swing—This is the peak output voltage range, referred to zero, that you can obtain without voltage clipping.

Current Offset—At very low impedance levels (1 k ohm or so), you can usually ignore the input currents and input offset current. If the impedance is higher, current may become important. When you're working with low impedance values, don't worry about anything except input offset voltage. However, when the impedance is greater, you'll have to take both voltage and current into consideration. Current problems are usually caused by a mismatch in the input stage. See Chapter 2, "Differential Input Impedance."

Slew Rate—This is the maximum rate of change of the output voltage with respect to time, that the **OP AMP** is capable of producing and still operate in the linear mode. You'll usually find it listed in *volts per microsecond*.

You'll find other terms that will help you when working with **OP AMP's**, listed and explained in Chapter 2. But it must be remembered that bandwidth, slew rate, output voltage swing, output current, and output power of an **OP AMP** are all interrelated. Furthermore, these characteristics are frequency and temperature-dependent, as has been previously explained. Because of all these problems, you'll find circuits for reducing offset current, frequency compensation, and bias current, are included internally and externally with every **OP AMP**.

Voltage Gain Measurement

Basically, a voltage gain measurement for an **OP AMP** is the same as for an audio amplifier (as discussed in Chapter 1). Just remember, don't overdrive any **IC, OP AMP**. It's possible to damage the **IC** and, almost without question, you'll end up with erroneous readings. The best approach to testing one is to apply the manufacturer's maximum rated input, then measure the output. If possible, check the output with a scope. If you see clipping taking place when you apply the recommended input signal (before you make a measurement), reduce the input signal until the clipping stops.

Next, increase the input signal frequency until the voltage gain drops 3 dB from the initial low frequency value. This is the **OP AMP's** *open-loop bandwidth* and voltage gain (see Chapter 2). To check the *closed-loop gain,* use the same procedure, except now the **IC** must have a feedback circuit. The closed-loop gain of an **OP AMP** is dependent upon the ratio of the feedback and input resistance (see

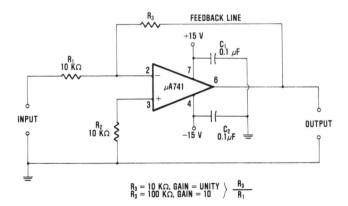

$$R_3 = 10 \text{ K}\Omega, \text{ GAIN} = \text{UNITY} \;\big\rangle\; \frac{R_3}{R_1}$$
$$R_3 = 100 \text{ K}\Omega, \text{ GAIN} = 10$$

Figure 5-3: A typical example of a feedback OP AMP operating in the linear (inverting) mode. The gain of such a circuit is determined by the ratio of external feedback resistance. Circuits may differ from the one shown by having more components, but the basic operation is still the same

Figure 5-3). When you check an **OP AMP's** closed-loop voltage gain and bandwidth, you'll find lower readings than your open-loop measurement values. In other words, the frequency response won't be as good and the gain will be less.

Input/Output Impedance Measurement

As we explained in the preceding section, when you measure the characteristics of an **OP AMP**, it must be remembered that the open-loop values will be higher than closed-loop values. This also is true of impedance values, i.e., open-loop impedance will differ from closed-loop impedance. To find the input impedance with the **OP AMP** operating into a load, construct a test setup like the one shown in Figure 5-4. Incidentally, if you're checking the same **OP AMP** as the one used in the preceding measurement (voltage gain), be sure that you use the same frequency, or frequencies, load, and test gear for this measurement as you did for that test.

After you have the setup shown in Figure 5-4, adjust the sine-wave generator to the frequency at which the **OP AMP** is to be operated. Now, adjust your variable resistor (shown as R in the drawing) until you read exactly the same signal voltage on each of the voltmeters. Finally disconnect the resistor and use the ohmmeter to measure the resistance between resistor terminals 2 and 3. What you read with your ohmmeter is the dynamic input impedance of the **OP AMP** under test. *Note:* if the test frequency is very high, be sure to use a non-inductive load resistor. Another point: obviously, you can use one voltmeter in place of the two and, to speed things up, a resistor substitution box can be used in place of a single variable

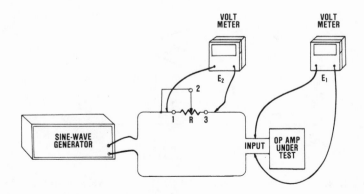

Figure 5-4:   Test setup for measuring the input impedance of an OP AMP
under load

resistor. But don't forget, the accuracy of the measurement will depend on the tolerance of the resistor used as well as the accuracy of the ohmmeter.

To measure the dynamic/output impedance of an **OP AMP**, use the measurement test described in Chapter 2, under "How to Make a Dynamic Load Test." The connections and procedures (see Figure 2-13) are the same. The load resistor is varied until you have maximum power being delivered to the load. Then power is disrupted and the load resistor removed from the test circuit. Measure the value of the load resistance with the ohmmeter. Whatever you read is equal to the *dynamic output impedance* at the test frequency you used to make the measurement.

Distortion Measurement

As you are probably beginning to realize, in almost every case, **OP AMP** measurements are just the same as you'd make in any audio amplifier, and this measurement isn't any different from the rest. The procedure for an effective linear **IC** amplifier distortion test is given in Chapter 2. (See Figure 2-10 for the test setup using an oscilloscope and square-wave signal generator.) Figure 2-11 shows the various scope patterns of a distorted square wave that could be observed during testing of an **OP AMP** or similar device.

Noise Measurement

Noise measurements for **OP AMP's** are the same as for audio amplifiers. However, as with any audio amplifier, the frequency spectrum or random noise will show different output levels at different frequencies, and is dependent upon the amplifier under test. By referring to Figure 5-5, you'll notice that (A) shows the voltage amplitude constant from 0 Hz to 10 kHz and beyond. This is called *white noise*. You may see this type noise referred to as *Johnson noise* but, in the strictest sense, Johnson noise is the noise generated by any resistor at a temperature of absolute zero in degrees Kelvin.

In practice, a perfect noise output level isn't very common because an amplifier will affect the output at various frequencies. For example, in Figure 5-5 (B) the amplifier has higher gain at the low frequencies than at the highs. This type of noise spectrum is called *red noise*. On the other hand, if the amplifier has better gain in the high frequency region, you'll find the output graph would look like (C). This is called *blue noise*. Next, the graph in (D) shows what is called *pink noise*.

Typical noise for a low-cost **OP AMP** such as the 741, when

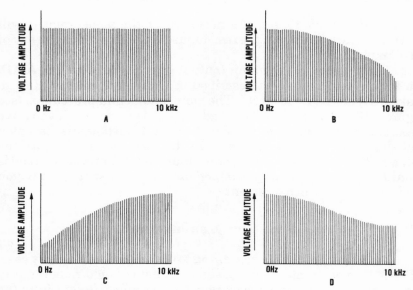

Figure 5-5: Graphs of an audio amplifier's effect on white noise spectra. (A) white noise, (B) red noise, (C) blue noise and (D) pink noise

referred to the input, is 10 μV, which isn't particularly good. Thus, if your **OP AMP** has a gain of 10, you should measure 100 μV on the output. At a gain of 100, you'd have 1 millivolt out, and so on. What this means is that you will need a pretty good scope to make a noise measurement (see Chapter 9). The test setup is shown in Figure 5-6.

To make the measurement, adjust the gain control on your scope until you have no noise showing. Your scope should be showing a light line across the face. Next, increase the scope gain control until you have a measurable peak-to-peak amplitude. Now, read the peak-to-peak amplitude of the noise off the face of the scope. *Note:* it's possible to get an erroneous reading due to the scope leads picking up AC hum. To check this, set your scope sync control to operate at 60 Hz. If you can get the pattern to appear stationary, you're picking up 60 Hz interference. Also, **OP AMP's** can break into oscillation. If you suspect this is happening, short the input to the **OP AMP** circuit. If you still see hash (sometimes called *hum*), the **OP AMP** circuit, more than likely, is oscillating.

How to Measure Input-Offset Voltage and Current

DC base bias current *must* be provided for both the + and – input to an operational amplifier. In theory, the input bias current

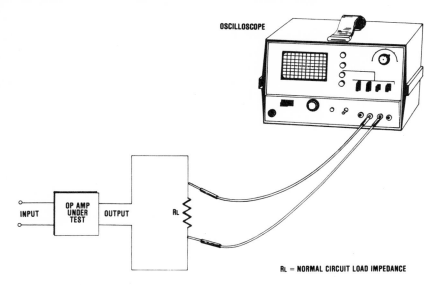

Figure 5-6:   Test setup for measuring an OP AMP's noise level. The scope should be capable of a measurable deflection with 1 millivolt or less. Generally, this measurement is made with the OP AMP operating in the open-loop mode.

should be the same for both inputs. In reality, they should be as close to equal as possible. However, even with the best of matching, there will be a slight difference between the two. Of course, this will cause a slight voltage difference between this offset voltage and true input signal. A handy rule to remember is: the output offset voltage will equal the input offset voltage times the in-circuit gain of the **OP AMP**. The test setup is shown in Figure 5-7 (page 114).

    As a typical example, let's assume that you have a test setup using these resistor values: $R_1 = 51$ ohms, $R_2 = 5.1$ k ohms and $R_3 = 100$ k ohms. Next, under these conditions, we'll assume that you measure the two voltage outputs and find them to be $E_1 = 80$ millivolts (jumper leads connected) and $E_2 = 360$ millivolts (jumper leads removed). All you need now are two simple calculations:

$$\text{input-offset voltage} = \frac{E_1}{\left(\dfrac{R_2}{R_1}\right)} = \frac{80 \text{ mV}}{\left(\dfrac{5.1 \text{ k ohms}}{51 \text{ ohms}}\right)} = 80 \text{ mV}/100 = 0.8 \text{ mV}$$

$$\text{input-offset current} = \frac{E_2 - E_1}{R_3\left(1 + \dfrac{5.1 \text{ k ohm}}{51 \text{ ohm}}\right)} = \frac{280 \text{ mV}}{100 \text{ k} (1 + 100)} = 27.7 \text{ nano A}$$

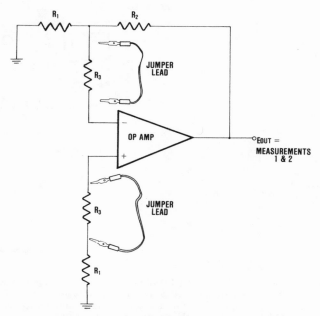

Figure 5-7:   Test setup for measuring an OP AMP's input-offset voltage and current

Using a Square Wave to Measure Slew Rate

The slew rate problem of an **OP AMP** can be stated in one sentence. The slew rate of an operational amplifier limits large output, high frequency signals. Or, to put it another way, ordinarily, you can't have large output swings and high frequency operation at the same time. Perhaps, the easiest way to observe and measure the slew rate of an **OP AMP** is to measure the slope of the output waveform of a square wave input signal, as shown in Figure 5-8. When you're making this check, it's important to remember that your square-wave generator *must* have a rise time better than the slew rate capability of the **OP AMP** you're testing. For example, the fastest you can normally change the output of a 741 is 0.5 volts per microsecond. Therefore, as long as your square-wave generator has a faster rise time, your measurements will be correct because any distortion you see on the scope will be caused by the **OP AMP**, not the test equipment.

Troubleshooting OP AMP's with a Scope

When you hear servicemen speaking of troubleshooting today's solid state equipment, more than likely you'll hear at least

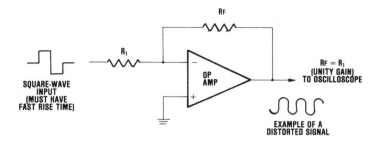

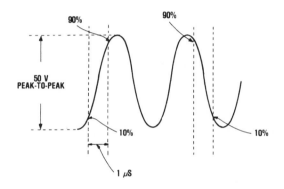

EXAMPLE SHOWS A SLEW RATE OF APPROXIMATELY 50 (50 V/μS) AT UNITY GAIN.

Figure 5-8:   Test setup and example slew rate measurement pattern that
might be seen on a scope when checking an IC such as an LM
318

one of them say, "The oscilloscope is the most valuable tool in the
field of electronics measurement." While this may be an
overstatement, it's so close to accurate that it's surprising to find so
many technicians using a scope without understanding how to
troubleshoot with it effectively. Perhaps one of the reasons for this is
that the price of high-performance scopes has decreased
dramatically. For about $300.00, the service technician can
purchase a 10 kHz triggered oscilloscope with a vertical input
sensitivity of 10 millivolts per centimeter. Suddenly, we find
ourselves face-to-face with items such as x 5 sweep magnifiers, auto
and normal trigger level controls, and a host of other specialized
functions that, because of cost, didn't even exist for most of us just a
few years ago. But once you learn to operate the scope, these controls
become an asset. On the other hand, if you don't follow some simple
rules, troubleshooting **OP AMP's** with a scope can be troublesome.

As has been explained in previous chapters, a wide variety of useful tests on audio equipment using **OP AMPs** can be made with a scope. It is a general rule that your test equipment must have performance characteristics equal to, or better than, the **OP AMP** (or any other device) under test. There are, however, certain exceptions that are possible if you use the right test procedure. For example, amplitude nonlinearity is a fundamental cause of distortion in an **OP AMP** audio amplifier.

If you want to make a linearity check with a scope—using the following procedure—you must first determine the linearity of the scope itself. Large percentages of distortion in any audio waveform are easy to "see," but you really have to look *very* closely to see small percentages of distortion such as illustrated in Figure 5-9. If you have good eyes, it's possible to estimate the percent of distortion by

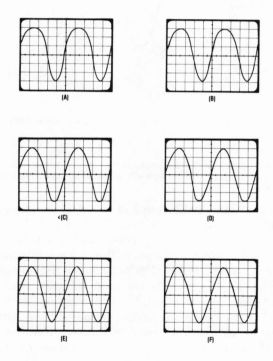

Figure 5-9:   These examples clearly show how difficult it is to see various percentages of distortion, using a scope. (A) 20%, (B) 15%, (C) 10%, (D) 5%, (E) 3% and (F) 1%

viewing a sine-wave test pattern on your scope, as shown in Figure 5-9. A good way to improve the effectiveness of a scope measurement is by utilizing Lissajous patterns.

Checking the scope's amplifiers, provides you with a reference pattern for use in evaluating the linearity of the audio amplifier under test. To make this check, connect the output from an audio oscillator to both the vertical and horizontal inputs of your scope, as shown in Figure 5-10.

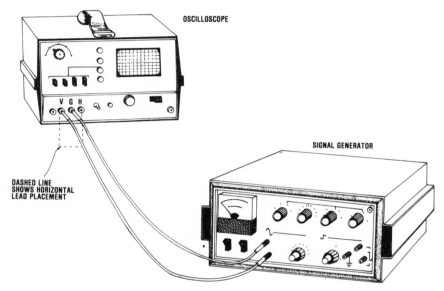

Figure 5-10: Test setup for making a linearity test of an oscilloscope. This test setup provides a reference pattern for testing an OP AMP, etc

As you can see by referring to Figure 5-10, you should connect the output from your audio signal generator to both vertical and horizontal input terminals of the scope. Don't worry about the linearity of the audio signal source you're using because the waveform presented at its output is of no concern when using this procedure. Next, set the audio signal source operating frequency to approximately 400 Hz. Now, you should see a diagonal line displayed on your scope screen. If your scope amplifiers are linear, you'll see a perfectly straight line, but if the scope is introducing distortion, the line will be slightly bent; i.e., show some curvature, as in Figure 5-11. The perfectly straight line is what you want for a reference, to make an accurate evaluation of the amplifier under test.

The rest of this test is basically the same as that for the square-wave testing of **IC's** explained in Chapter 2. The test setup for using

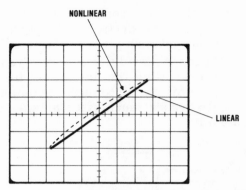

Figure 5-11:   Scope presentation of a reference linearity pattern showing a slight curvature. The curved line indicates the scope's amplifiers are non-linear

this procedure is shown in Figure 5-12 and, as in square-wave testing (see Figure 2-10), the load resistor, R, should be equal to the recommended load impedance for the **OP AMP**. Generally, when testing a low power **OP AMP**, you don't have to worry about the wattage rating of the load resistor.

Finally, observe the pattern on the scope screen. If it is exactly the same as your reference pattern, the **OP AMP** under test is linear. If you make a low power output measurement and a maximum rated power output measurement, you'll usually find that the linearity improves as you reduce power out. *Note:* it may be impossible to see any departure from your reference line if the **OP AMP** has good performance characteristics.

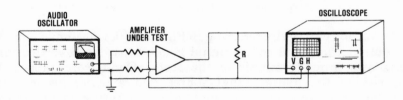

Figure 5-12:   Test setup for checking OP AMP linearity, using a scope

Most of us tend to assume that an oscilloscope display is telling us what actually is happening in a circuit under test. Believe me (and I'm speaking from experience), this may or may not be true! Select the wrong scope probe and it's very possible that you'll see some strange things during certain tests. Take a look at Figure 5-13. You probably can tell, by close examination, that it's a weird-looking

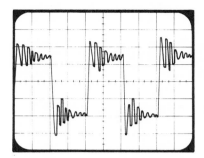

Figure 5-13:   An example of overshoot and ringing, caused by using the wrong probe/scope combination.

square-wave pattern. What's the trouble? It's simply a case of someone (me, in this case) connecting the wrong probe to a scope. What has happened is that a service-type C probe was connected to a high-performance scope. The test signal fed through the service probe to the scope was a 1 MHz square wave with a rise time of 0.02 μ sec. What you're seeing is a large amount of overshooting and ringing. The moral of the story is, use a good (lab-type) probe with a good (lab-type) scope.

Unless you take certain precautions, especially at high frequencies, getting the signal to your scope can be the toughest job encountered when making a test. For example, if you fail to match the probe to the scope, it can result in as much as a 50% measurement error. Fortunately, there are three rules that will help, under most testing conditions. These are:

1.   If loading error is a problem, you can reduce your error to less than 1% by selecting a scope/probe combination that has a resistance, looking into the probe, of at least 100 times as great as the signal source impedance.

2.   To keep your frequency-related errors down, select a scope/ probe combination with a shunt capacitance as small as possible.

3.   To reduce phase measurement errors, use low impedance scope/probe combinations.

Always use a low-capacitance probe instead of a direct probe, in order to avoid pattern distortion caused by circuit loading. If you're checking signals in high-frequency circuits, a demodulator probe should be used. But don't make the mistake of using a demodulator probe when a low-capacitance probe should be used. For example, don't use a demodulator probe when checking a TV set in the video amplifier circuits (anywhere after the IF stages).

OP AMP Regulator Circuit Troubleshooting

An operational amplifier is one of the nicest things that has happened to DC power supply regulators over the years, because such amplifiers have a differential input stage and all stages are DC

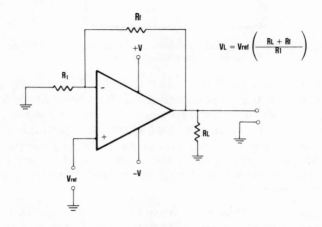

$$VL = Vref \left(\frac{RL + Rf}{R1} \right)$$

Figure 5-14: Basic idea of how an OP AMP is used as a voltage regulator

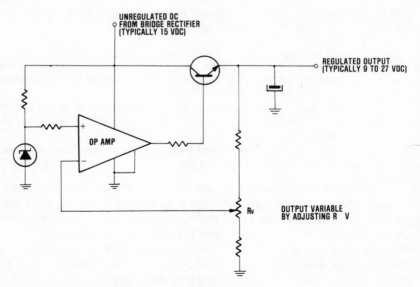

Figure 5-15: Voltage regulator using an OP AMP

coupled. Another thing in their favor is their open-loop gain. The voltage regulator shown in Figure 5-14 illustrates the basic idea involved in using an **OP AMP** as a regulator.

In practice, you'll find that the reference voltage usually is provided by a zener diode and the output voltage often is made variable by including an adjustable resistor in the input lead of the **OP AMP.** Another addition is an emitter follower on the **OP AMP's** output lead. The purpose of the emitter follower is to increase the current output capability of the regulator (see Figure 5-15).

CHAPTER **6**

Simplified Testing and Servicing of

Solid State Oscillators and

Waveshaping Circuits

The real joy of electronic equipment servicing comes from being able to troubleshoot tough jobs and working with state-of-the-art equipment. With this chapter, no matter which kind of audio oscillator or waveshaping circuit shows up on your workbench, you'll be able to easily and effectively repair complicated malfunctions in less time than you thought possible!

Oscillators and waveshaping circuits represent a growing area of electronic technology; for example, electronic music systems. The modern electrophonic music systems utilize oscillators, wave filters, and waveshapers that, in some respects, can be compared to electronic organ circuits.

In the past, we all used the terms "oscillator" and "signal generator" to describe the same piece of gear. Today, you'll find *oscillator* and *signal generator* indicate different technologies used to produce the fundamental signal. As an example, the master oscillator circuit of an electronic organ is designed to have a natural resonance and able to produce a pure sinusoidal signal. As you can see, the term "oscillator" is applied to this type of circuit. On the other hand, the term "generator" normally indicates some other form of electronic circuitry; for instance, an electrophonic music system designed to produce novel musical forms that do not necessarily follow any established rules. In fact, what comes out quite often is determined by a random number or random noise source. Incidentally, referring back to electronic organs, the output of their master oscillators (they have quite a few) are fed to **IC** frequency dividers and, when you include these devices, the circuit is called a *tone generator*.

Sine-Wave Oscillator Review

At the moment, we are primarily interested in sine-wave oscillators. Although Figure 2-16 showed the three most frequently used basic audio oscillator **IC** circuits, for more detailed explanation, Figure 6-1 shows them again.

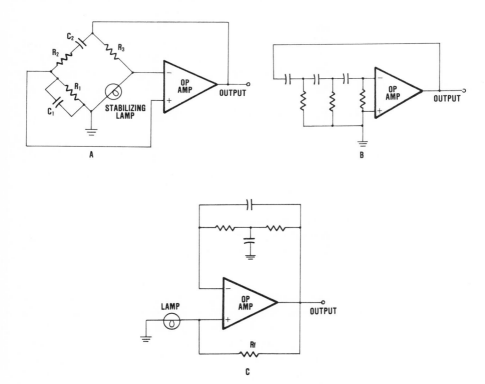

Figure 6-1:   Three of the most widely used modern sine-wave oscillators, (A) Wien bridge, (B) phase shift, (C) bridge-T

Each of the circuits has attributes that can be used to an advantage in certain situations, as was explained in Chapter 2. Probably the Wien bridge oscillator circuit shown at (A) is the best known to servicemen. It's been around since the days of vacuum tubes and just about the only difference is the operational amplifier IC.

There is one basic similarity among all oscillators; they must have positive feedback to sustain oscillations. Using an **OP AMP** for an oscillator is a natural because they have a ready tendency to oscillate. In fact, if circuit phase shift causes the feedback to become in-phase with the input, and the loop gain is high enough, you'll end up with an oscillator rather than an amplifier, no matter what you're trying to do. The process of stopping an **OP AMP** from oscillating is called "compensation." However, that's not what we're interested in, at the moment.

At this point you may be wondering, "How does one change the frequency of a circuit like this?" The answer is, "Easy." For example, the frequency of both the Wien bridge (A) and phase shift (B) in Figure 6-1, is directly dependent on the value of capacitance you select. A 10 to 1 change in capacitance value causes a 10 to 1 change in frequency (all other components remaining unchanged). This, of course, is one reason why these designs are in such wide use.

Guide to Servicing a Wien Bridge Oscillator

As was mentioned before, basically, any oscillator is constructed using an amplifier and a feedback network. Figure 6-2 shows a simple block diagram of an oscillator with signal phase requirements. The primary requirements are that the output signal of the amplifier must be larger than the input signal and inverted 180°. The output of the amplifier is fed through a phase-shifting network, which changes the phase another 180°, making it *in phase* with the input of the amplifier. If these requirements are met, and assuming the signals are the same amplitude, the circuit will sustain oscillations.

In real life, some experimenting with a circuit such as the one shown in Figure 6-2 would be required. In order to achieve stable oscillations, you would probably have to adjust the gain of the amplifier and make experimental changes in the feedback and biasing circuits. Figure 6-3 is a diagram of a Wien bridge oscillator. The schematic of the Wien bridge shown in Figure 6-1 (A) is identical in operation. The only difference is the way the components are laid out in the drawing. The purpose of the difference is to make it a little easier to understand the operation. The Wien network, composed of R_1-C_1 and R_2-C_2, provides the positive feedback path, and R_3 and the

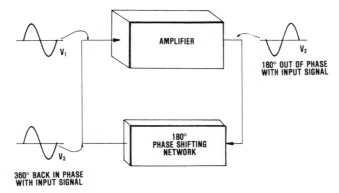

Figure 6-2: Block diagram showing principal requirements of an oscillator feedback system

stabilizing lamp provide the negative feedback. If either feedback is greater than the other, oscillations will die out, so it's necessary to provide some means of automatically controlling the feedback. That's where the lamp (other devices such as thermistors, nonlinear resistors, zener diodes, FET's, etc., are used) comes in. In this case, the lamp regulates the amount of negative feedback.

If you're working with a commercial audio oscillator using a Wien bridge, you'll probably find it is continuously tuned by a variable capacitor, and the ranges usually are changed by changes

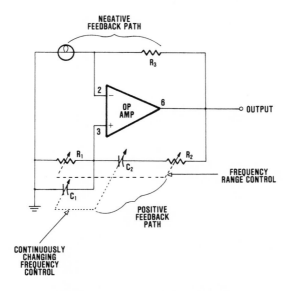

Figure 6-3: Wien bridge oscillator operational diagram

in resistance. (See Figure 6-3 for an example using these components (R_1-C_1, R_2-C_2) to change frequency.)

When troubleshooting a Wien bridge oscillator of the type shown you should find about 20 V peak-to-peak on the output of the **IC**. Next, if you set R_1 to equal R_2, and C_1 to equal C_2, you can then use the formula $1/(2\pi R_1 C_1)$ to determine the frequency output. For proper operation, it's essential the **IC** gain be sufficient (large at the frequency of operation) and that the oscillator operate with a net phase shift of the two RC combinations of zero. If you're having trouble with amplitude variation, check the tungsten lamp (although they are not as popular, you may find a thermistor, etc., being used in place of the lamp).

How to Troubleshoot a Phase-Shift Oscillator

The phase-shift oscillator, like the Wien bridge, is also an RC circuit with about the same frequency tuning range (a few hertz to 10 MHz). To see how this oscillator phase-shift circuit works, consider the network in Figure 6-4. If you connect a signal source (V_s) to the network input, as shown, you'll find that at a certain frequency (determined by R and C), the signal voltage across R_1 will lead the voltage output of the signal source by 60°.

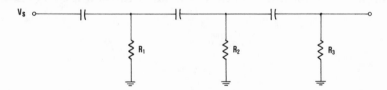

Figure 6-4:   Phase-shift oscillator, phase-shift circuit (see text)

Next, check the voltage across R_2. It will lead the input signal by 120° (assuming all capacitors and resistors are of equal value). The voltage on the final resistor (R_3) will lead by 180°. Now, remembering that we get a 180° phase shift by passing a signal through a common emitter amplifier, it becomes apparent that by passing the transistor output voltage through the phase-shifting network and then back to the transistor amplifier input, we end up with an in-phase feedback. The phase-shift oscillator in Figure 6-1 (B) uses an **IC OP AMP** as an amplifier, however, the basic operation is as just described.

Most electronics servicemen consider an oscilloscope second in importance only to a basic transistor voltmeter (TVM). You'll find that a scope can be used to check the gain of the solid state device as well as an excellent phase measuring instrument. Also, a calibrated vertical amplifier permits you to quickly measure the peak-to-peak voltage of a displayed waveform. When you select a scope to troubleshoot an oscillator circuit (or any other circuit), just remember one important point: The scope used in high-fidelity troubleshooting procedures should have better characteristics than the oscillator under test.

DC voltage and resistance measurements are basic in oscillator troubleshooting procedures. You'll find normal voltage and resistance values are specified in service data. If you note any changes in DC voltage and/or resistance values, it is probably an indication of circuit malfunctions.

Many modern TVM's are provided with a low power ohmmeter function in addition to a conventional ohmmeter function. There is an advantage when you use the low power resistance measuring function in that it applies a test voltage of less than 0.1 V to the points being measured. Now, because solid state devices normally need more than 0.1 V to start them conducting (0.2 V for germanium transistors and 0.6 V for silicon type), many resistors in solid state circuitry can be checked without removing them from the PC board, etc. However, it should be remembered that a defective capacitor, diode, or other solid state device can cause inaccurate resistance measurements.

Troubleshooting Bridged-T Oscillators

One of the most popular uses of the bridged-T circuit is to generate fundamental signals in audio oscillator test instruments. Usually, these instruments are continuously tuned by varying one resistor (R), and the range is changed by the selection of fixed capacitor (C) (see Figure 6-1(C). Incidentally, it also is possible that you'll encounter audio oscillators that vary the value of the capacitors and use fixed resistance. Either way, the troubleshooting procedure is about the same.

If you're working with a bridge-T (RC) audio oscillator test instrument, it frequently is designed to work into some certain impedance (for example, 600 ohms). Although the oscillator will operate properly into this impedance, it's quite possible that the internal output impedance will be much lower. Therefore, in some cases, you may have to place a resistance equal to the difference between the oscillator impedance and the load impedance, in series with the oscillator output, to bring about a proper impedance match; i.e., to increase the oscillator output impedance until it matches the load impedance. The test procedures for operational amplifiers used in low frequency sine-wave oscillator circuits are essentially the same as for other **IC** audio amplifiers (see Chapter 2 under "Troubleshooting **IC** Oscillators").

Testing Solid State Crystal-Controlled Oscillators

Transistors operate efficiently in crystal-controlled oscillator circuits such as the Pierce shown in Figure 6-5. This oscillator is very popular because of its simplicity and low cost of construction (it requires a minimum of components). Notice that a parallel inductor and capacitor are not required to form a tuned circuit for frequency control. The crystal is the only tuned circuit used to control the frequency. Although the circuit shown in Figure 6-5 is inexpensive to build and works well, the new wrinkle is to use **IC's**. Figure 6-6 shows the connections for a crystal-controlled rf oscillator built around an **IC**.

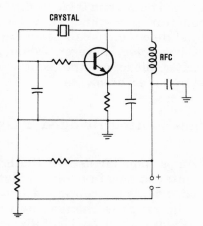

Figure 6-5:   Basic crystal-controlled oscillator (Pierce)

When troubleshooting an rf oscillator such as the one shown in Figure 6-6, a good place to start is, of course, to check the **IC** voltages first. Next, notice that a negative feedback loop is provided. The correct feedback voltage is developed by the voltage divider R_1 and R_3, and this voltage is applied to the inverting input terminal of the **IC**. Check these resistor values because if they change value, chances are that the oscillator will stop working. If this happens, try changing R_1 and/or R_3 resistance value until you get the oscillator going again. The actual feedback voltage is developed across R_1. Incidentally, the oscillator should work without any negative feedback. In fact, you may find some circuits designed this way,

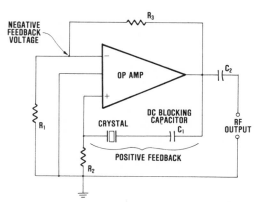

Figure 6-6:   Crystal-controlled rf oscillator designed to use a multi-stage IC

especially if a high harmonic output is desired. Also, if you're servicing equipment that uses overtone crystals, they do not oscillate at any *exact* submultiple of the marked frequency. You'll find them to be close, but not exact.

If you suspect the crystal is bad in an rf oscillator, it may be possible to substitute a coil and capacitor wired in parallel (an rf tank circuit) for the crystal. For example, remove the variable capacitor and loopstick coil from any old broadcast receiver you happen to have around the shop and use it in place of the crystal to produce a frequency that falls in the broadcast band. Simply plug the circuit into the crystal holder and then fire up the oscillator. Listen on any broadcast receiver and you should hear a squeal when you zero-beat against a broadcast station. This trick may not work every time (sometimes it may be all but impossible to get the oscillator started). You'll find the frequency will shift but, if you can get an output, it's a pretty good indication that the crystal you're checking is bad.

How to Service Modern Square-Wave Generators

Many test instruments (especially audio oscillators) include—in addition to a sine-wave output—a square-wave output. In general, you'll find that the output signal specifications are about the same as the ones for the sine-wave output. When you read the manufacturer's specs, the output voltage listed is usually the maximum voltage the oscillator can produce, and the output impedance is what is measured at the output terminals with the load disconnected.

Another term you'll encounter is *output amplitude symmetry*. This tells you how close to perfect you should find the square wave. In other words, how close to equal are the positive and negative portions of the square wave? Theoretically, they should be *exactly* equal. However, in practice, they will vary from a few millivolts to quite a few millivolts. In all cases, you'll also find that the widths of the negative and positive portions of the square wave will vary. When checking the width, you'll probably find that, under the best of conditions, about ±5% is the best you can expect.

The reason for pointing out the changes in a square-wave shape is that you can expect to find various square-wave generators where it is common practice to produce a square-wave output by using a sine-wave input and saturating a high-gain amplifier. Figure 6-7 will give you an idea of how this may be done, using an **OP AMP IC**.

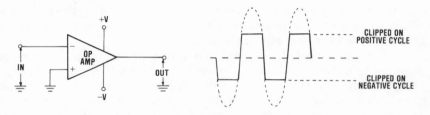

Figure 6-7:   Example of a sine wave being converted to a square wave, using an OP AMP IC

If you're checking an audio oscillator for distortion, it's extremely important that you use full-shielded cable with BNC connectors. If you don't, you're sure to run into hum and noise problems, particularly if you're trying for ultra-low-distortion. A further advantage is that using 50-ohm BNC connectors when working with square-wave signals, will help reduce reflection caused by impedance mismatches in your test setup.

If you try to produce a square wave using an **OP AMP**, as shown in Figure 6-7, there are a couple of precautions: 1) your input sine wave signal probably will have to be fairly large so, for safety, it's best to place a resistor in series with the **OP AMP** input terminal to prevent damage to the **OP AMP** input circuits; and 2) the **OP AMP** should be operating in the open-loop configuration (without feedback) so the amplifier can produce as much gain as possible.

One of the most common square-wave generators is the multivibrator. Figure 6-8 shows the connections for a simple **IC** multivibrator that can produce a fairly good square wave output signal. The circuit can be used with almost any **OP AMP** having a differential input circuit.

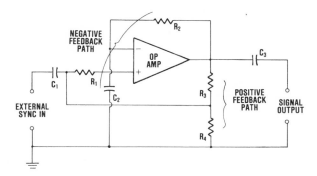

Figure 6-8:   Square wave IC multivibrator

Like all **OP AMP** circuits, if the feedback is improper, the multivibrator won't work. R_2 and C_2, in Figure 6-8, are the negative feedback circuit. However, the values of these two components also determine the oscillator frequency. If you suspect either of these two have changed value, one way to find out is to check the frequency output and use the formula, freq = $1 / (2 \pi R_2 C_2)$. The frequency will be approximate, and your answer will be in Hertz.

When troubleshooting a circuit like the one shown in Figure 6-8, you should expect to see the square-wave amplitude change, if you use a different **OP AMP** from what was in the circuit originally. The reason is that the amplitude of the square wave depends on the DC voltage being applied to the **IC** you are using. *Note:* you may or may not find a sync voltage being applied to the circuit, depending on its use. As we all know, a sync signal will make the difference between a very stable output signal and an unstable output. Incidentally, you'll also find that a sync signal is required if the circuit is being used as a frequency divider network.

Figure 6-9 is a working schematic of a transistorized multivibrator that will produce a square wave output signal. If you're working with a circuit similar to this one, one of the first things to notice is the resistance ratio between the collector resistor (R_C) and the emitter resistor (R_E). If you find the emitter resistor is about one-half the value of the collector resistor, the multivibrator is a high current type. This means that you must be careful of the transistor characteristics when choosing a replacement transistor. The main point is that the new transistor must have a high current capability and, of course, a high power dissipation rating.

On the other hand, if you find the collector resistor value to be about ten times larger than the emitter resistor, you're working with a low-current multivibrator. In this case, you don't have to worry so much about the power rating and current carrying capacity of the transistor. But, the low-current type is not as stable as the high-

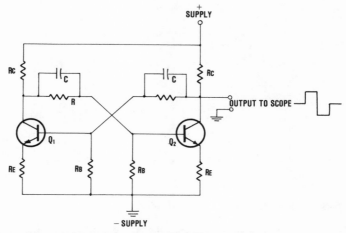

Figure 6-9:   Troubleshooting schematic of a multivibrator

current type, and the DC voltage power supply becomes much more important. If you're having trouble holding the output frequency constant, check the DC power supply, because any change in DC voltage amplitude will cause a change in the output frequency of the multivibrator.

Connect a scope to the output as shown, and you should measure a peak-to-peak voltage just a little over one-half the DC supply voltage (if you're checking a high-current multivibrator). However, the lowest collector voltage should be about one-third of the collector voltage and the highest, about two-thirds or more (but not equal to the supply voltage).

If you can't get a multivibrator started, an old trick is to place just about any solid state diode you have on hand in either of the feedback circuits (see Figure 6-9). However, if you're monitoring the output with a scope, you'll find the output waveform is asymmetrical until the multivibrator reaches full operation, which may be undesirable in some applications.

Another major advantage of these circuits is that very little current is drawn between pulses. Therefore, high current can be drawn from the output for short periods of time without damaging the transistor (assuming the pulse repetition rate is not extremely high).

Figure 6-10 is the working schematic of a unijunction transistor (**UJT**) multivibrator that employes fewer components than the one shown in Figure 6-9. When the circuit is operating, you'll find that the **UJT** is cut off during the charging cycle. During this time, capacitor C_1 is charged through resistor R_2 and diode CR_1.

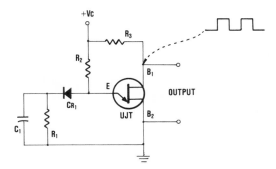

Figure 6-10:   Free-running UJT multivibrator working schematic

With a scope connected to the output, you should be able to measure the time period, which is determined by the time constant of C_1 and R_1.

Generally, the current through the transistor and the associated power dissipation are tough to calculate when you're dealing with multivibrators. Therefore, it is suggested that you use an exact replacement transistor, or one that is rated for a higher current and power dissipation when operated at the desired frequency. You'll find the frequency to be approximately equal to the reciprocal of the time constant.

Checking a Blocking Oscillator

Figure 6-11 is the working schematic of a blocking oscillator. The blocking oscillator is especially attractive because it can be triggered to produce short duration, high amplitude pulses. A pulse rise time of $0.1\,\mu$ sec is common. Feedback is accomplished by use of a transformer that drives the transistor into heavy conduction (in fact, into saturation). The transformer also is driven into saturation, which makes it difficult to change the shape and size of the pulse without changing transformers. Although it is possible to use any transformer with a blocking oscillator, the primary must be able to withstand the voltage and current, and the secondary must have the correct output impedance. But, as shown in Figure 6-11, often you'll find special purpose transformers used with blocking oscillators. In this circuit, R and C determine the resting period between pulses.

Normally, you will not find a blocking oscillator used in any application operating at radio frequencies, or in circuits designed to produce a sine wave. A typical output waveform is shown in Figure 6-12. Notice, very little current is drawn between pulses. The peak

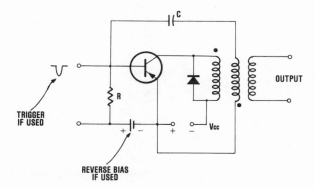

Figure 6-11:   Blocking oscillator circuit. If the circuit is triggered, it requires reverse bias on the transistor. Without reverse bias, the circuit is free-running

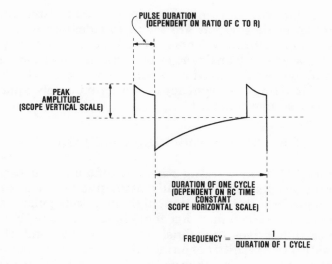

Figure 6-12:   Typical scope presentation of a blocking oscillator output waveform that may be used to check waveform amplitude, and frequency

output voltage (pulse) depends on the transformer turns ratio, but it may be made approximately equal to the supply voltage by adjusting the output-to-input turns ratio to 1:1 (in many cases, a step-down transformer is used; usually 5:1).

You should find the reverse bias about one-half the peak trigger voltage (if used). On the other hand, if you use a scope and measure the pulse peak, you may find the collector voltage is about twice the supply voltage. Usually, these high voltage peaks are not desirable and the diode is placed in the circuit to reduce the output voltage. The diode can be placed across the entire transformer primary winding, or only one-half, depending on the transformer construction (center-tapped or single coil primary).

Triggered blocking oscillators are fairly stable at their operating frequency, so far as DC power supply voltage variations are concerned. In the free-running mode, frequency is primarily determined by the R and C values. But, don't depend on most of the formulas given (f = $1/2\pi RC$, etc.) for calculating the frequency of free-running relaxation type oscillators because the transformer characteristics can affect frequency output. Your best bet, if the circuit is operating out-of-bounds (amplitude, pulse duration, or frequency), is to observe the output on a scope and adjust R for a frequency correction, C for a pulse duration correction, and check the diode in case of overshoot (although you may have to adjust both R and C to get the pulse duration you want).

If you can't get the oscillator to trigger, check the reverse bias—it's probably too high. You also can get into trouble when the reverse bias is too low. In this case, you may see the circuit triggered at the wrong point in time, causing other signals to be mixed with the trigger waveform.

Converting Sine Waves to Rectangular Waves

It is often necessary to modify sine waves to achieve a certain test pattern; for example, in audio oscillator test instruments, where squaring circuits are used extensively. The Schmitt trigger is the circuit most commonly used to generate a square wave from the internal sine wave of the audio oscillator. Ordinarily you'll find that the Schmitt trigger circuits used in audio oscillators are designed to produce comparatively slow rise times. The idea is to attenuate noise as much as possible. Like other circuits of this type (Schmitt triggers are widely used in triggered sweep oscilloscopes), the input to the squaring circuit must be buffered so it does not affect the sine wave circuits by loading.

Typically, a Schmitt trigger converts an input waveform of any shape to a rectangular pulse of fixed amplitude. You'll find them used in monolithic **TTL** monostable multivibrator **IC's** such as Signetics, Digital 54/**TTL** series and pulse generators using transistorized circuit design. Today, there are numerous ways to produce a square wave; for instance by using **OP AMP IC's**. However, we have to troubleshoot the old with the new so let's first examine a transistorized Schmitt trigger (see Figure 6-13).

A typical transistorized Schmitt trigger (Figure 6-13) consists

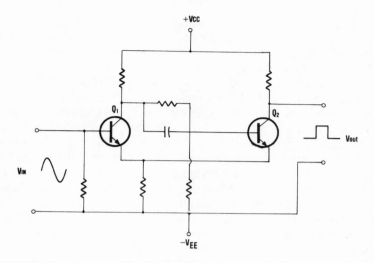

Figure 6-13:   Schmitt trigger circuit that can be used to convert an input
waveform of any shape to a rectangular pulse

of two emitter-coupled transistors. If you measure the voltage across
the collector resistor, you'll find transistor Q_2 in Figure 6-13 is
conducting and Q_1 is not (no voltage across its collector resistor).
This condition should remain constant until the input signal
amplitude overcomes the bias-emitter junction. As you can see,
troubleshooting a circuit like this is fairly easy. For example, remove
the input and the Schmitt trigger circuit should remain in a stable
condition. Reconnect it and the circuit should start its switching
action.

Triangle/Square-Wave Function Generator

A very useful circuit that is easy to construct is a triangle/
square-wave function generator, using two general purpose **OP
AMP's** such as Signetics μA 748 and a μA 741 (of course, you can
get these **OP AMP's** from several other companies such as
ANALOG, Fairchild, Motorola, etc.). As is apparent, (see Figure 6-14),
the circuit is fairly simple and straightforward.

The peak-to-peak output voltage amplitudes will depend on the
zener diodes used for D_1 and D_2. For example, the peak-to-peak
output voltage you should measure, if the zeners are 1N754's, is
about 15 V (E_{o1}, E_{o2}, and E_{o3}). However, be careful if you're using an
rms reading voltmeter to measure a function generator output
because it's necessary to use correction factors to determine true rms
voltage readings on this type instrument. Furthermore, different

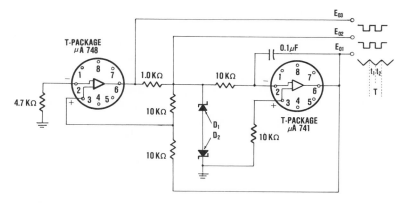

Figure 6-14:   Triangle-square-wave function generator constructed by using two general purpose OP AMP's

correction factors are required for different type rms calibrated voltmeters; for instance, average or peak responding types.

The time duration of one cycle can be determined by using the formula $2 \times$ peak-to-peak $(R_1 C_1)$/peak voltage. For example, with a peak-to-peak of 15 V, the peak voltage is 7.5 V. Next, Figure 6-14 shows R_1 to be 10 k ohms and C_1 to be 0.1 μF. So, placing these values in the formula, it comes out that the circuit shown will produce a waveform with a time duration very close to 4 milliseconds, or a frequency of about 250 Hz. This equation can be used to design the circuit shown in Figure 6-14 for a wide range of output voltages and output signal time periods (frequencies).

How to Troubleshoot Waveshaping Networks

I think we all agree that the only practical way to test waveshaping circuits is to use a scope. For example, the *rise* and *fall* times of a pulsed waveform indicate its steepness. Why do we care? Because the steeper the rise time, the more likely you'll have trouble with the leading edge overshooting and causing a ripple in the first part of the pulse top. Now, you'll find that many amplifiers will cause this form of distortion when you're working with pulse networks. Therefore, it's all but impossible to troubleshoot equipment using circuits of this type without knowing what the pulse *looked like before* it was applied to the amplifier. Without a scope, you can't make a comparison. The same applies to a waveforming circuit such as an oscillator. In either case, the exact *waveform, amplitude*, and *duration* can be measured directly with a scope. Furthermore, frequency can be found if the duration (or period) of one complete cycle (it's best to display two or three cycles on the scope to make the measurement) is measured, since frequency is the reciprocal of the waveform time period. By the way, this is only true provided you're not trying to check a waveform in the VHF radio frequency range. The reason for this is that many older oscilloscope amplifiers just can't make it at such high frequencies without altering the waveform of the signal under test. Even if the scope can pass the test frequency, it's usually very difficult to analyze a single high-frequency pulse. A digital counter is your best bet, in these cases. Figure 6-15 shows the basic scope connections you'll need for measuring low frequency waveforms.

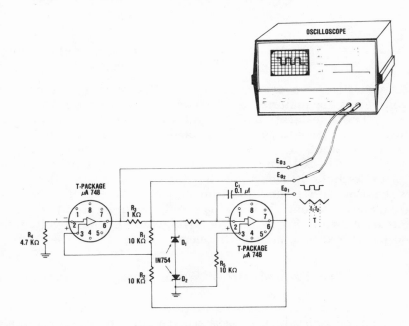

Figure 6-15:   Scope connections for measuring waveform, peak amplitude, and frequency

You'll find that there are some basic steps you can use to measure almost any kind of waveform that falls within the limitations of your scope. They are:

1. Set up the scope connections the same as the ones shown in Figure 6-15, if you're checking a circuit output (a good place to start). To check all phases of the waveform's geometry, your scope should be calibrated in both vertical and horizontal displays.

2. Apply power to the circuit under test.

3. Adjust the oscilloscope sweep frequency and sync controls

until you have two or three stationary waveforms displayed on the scope viewing screen.

4. Now you're ready to start making measurements. A good start is to measure the amplitude of the waveform under observation. To do this, use the voltage calibrated vertical axis of the scope. Of course, you must check the scope calibration first, but let's say that the scope is set for a sensitivity of 10 mV per cm. In other words, your scope would display a 10 mV peak-to-peak signal as 1 cm high on the CRT. But, watch it! Notice, we are talking about peak-to-peak, not rms. In fact, a 28.2 mV peak-to-peak sine wave voltage would read 10 mV on an rms reading voltmeter. A 10 mV rms signal would show up on the scope (sensitivity of 10 mV cm) as 2.82 divisions high because a 10 mV rms sine wave signal is 28.2 mV peak-to-peak signal. A simple solution to problems like this one is to stick to peak-to-peak voltages and you can't go wrong, regardless of the waveform under measurement.

5. To measure the duration of a waveform, you must use a calibrated sweep. You'll find that the majority of oscilloscopes use the centimeter as the basic horizontal division and low bandwidth scopes have a faster sweep speed of about 1 to 0.5 μ sec per division. The period of the time base is changed by adjusting the time base switch. Accuracy of the time base, typically, is 5% to 3%, when the scope is in perfect condition and under ideal test conditions.

6. To find the waveform repetition rate or output frequency, measure the duration of one complete cycle. *Note:* it is important to measure an *entire cycle*. Don't make the mistake of measuring the time of one alternation, i.e., one pulse, one-half a square wave, or other waveform. As an example, assume one complete cycle of a waveform under test occupies 10 horizontal divisions and the scope sweep speed is set at 0.5 μ sec per division. Under these conditions, one cycle is completed in 5 μ sec. Therefore, 1/0.000005, or 200 kHz.

If you're using a recurrent sweep (horizontal axes are frequency-calibrated), there is no way of making a calibrated time measurement except by comparison. If it is desired to find a waveform repetition rate or output frequency, set the scope sweep and sync controls to produce one complete cycle that fills the entire sweep length on the face of the CRT. Then read the frequency from the oscilloscope dials. The frequency of the scope sweep oscillator can be changed with a variable control and several switched steps, in most cases. A sweep frequency range from 5 Hz to 500 kHz is typical. This works out to be an equivalent time-per-division of 20 m sec per division to 200 m sec per division, assuming there are 10 horizontal divisions. However, these values are not useful (except under the roughest of conditions) as a standard for time comparison.

CHAPTER 7

Troubleshooting Late Model
Solid State Radio Receivers

In this chapter, localizing troubles in solid state radio receivers is explained in detail, and servicing techniques using test instruments are included throughout. The following pages not only explain receiver circuits and applications in *easily* understood terms, but also narrow in on troubleshooting procedures that really count—simplifying some and completely eliminating others. There are explanations of *how* and *why* each defect is causing the trouble, so you'll know exactly what steps to take to get your troubleshooting job off in the right direction—toward fast, accurate diagnosis and easy repair!

Analyzing Trouble Symptoms

Common sense tells us that the way to perform the most effective and profitable troubleshooting is to be accurate and fast. However, many servicemen lose time because they do it the slow, hard way. For example, you can approach diagnosing receiver troubles from two standpoints. First, a frequent approach, analyze a symptom, then localize the trouble by using a scope and demodulator or high impedance probe or, second—and I consider this the better way—in many cases, you can eliminate the extra time it takes to hook up a scope and its associated equipment. For instance, a bench timesaver that will localize faulty IF stages is to simply read the voltages across each of the emitter resistors of the IF stage transistors. You'll find that almost every transistor radio receiver uses a by-passed resistor in the emitter circuit. In general, you should measure a volt or two (DC). If you don't, more than likely the transistor is open. On the other hand, if your voltage reading is high, probably the transistor is shorted.

Accuracy requires that we be able to analyze trouble symptoms rapidly and *correctly*. In the days of tube receivers, about all the serviceman had to keep in his head was a general block diagram for the superheterodyne receiver. Today, he has to know much more. Sure, a mental block diagram of the set is a tremendous help at the outset, but there is a lot more about modern circuits that we must know if we're going to analyze trouble symptoms rapidly and correctly. There's no way around it. To get this technical data, most of the time we must consult the receiver service data. However, don't jump off into a whole lot of unnecessary work until you're sure it needs to be done. Don't pull a chassis before you make certain power is being supplied by the power cord. Check the antenna lead (if there is one); it may be broken. Look for a tripped circuit breaker and, if possible, check the DC supply voltage to see if the problem is in the DC power supply or receiver stages before you take any further action.

From time-to-time, we all suffer from an ailment called "Customer Fatigue." Generally, the symptom is a strong desire to ignore every word the customer has to say about his or her receiver problems. Although sometimes it's a very difficult task, encourage the customer to tell you what happened when the trouble started and whether it has been steady or intermittent. It's very possible that the trouble is nothing more than a loose connecting lead to the speaker. I ran into a similar trouble where opening and closing the customer's door turned the set off and on. In this case, it turned out to be a faulty power cord.

After you have completed a visual inspection and rattled the wires a bit, follow up with instrument tests to isolate the trouble to a specific section of the receiver. Next, make whatever additional tests are needed to localize the defective circuit. Then repair or replace the bad component and, finally, don't overlook a thorough checkout of the entire receiver before removing it from your workbench. This last step will save you a lot of time and unhappiness in the long run. Don't cut corners. Check everything!

Troubleshooting Antenna Circuits

The antenna and its associated circuits become very important to the serviceman when operation of the receiver must be maintained under very weak signal conditions. A good example is an auto radio where a small antenna is used. Because maximum receiver sensitivity is desirable for installations such as this, an antenna trimmer capacitor often is provided. See Figure 7-1.

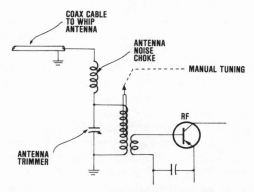

Figure 7-1:   The antenna system is most efficient when the trimmer capacitor is adjusted for peak receiver performance. Make sure there are no overhead wires near the whip antenna during trimmer adjustment.

Perhaps the least understood circuit operation in receiver servicing is the antenna system. For example, even with a perfect antenna, ideal testing conditions (nothing affecting the antenna environment), a simple mismatch between the transmission line (coax cable in Figure 7-1) and first amplifier input circuit, will reflect a portion of the incoming power back to the antenna, where it is *re-radiated back into the atmosphere* and may be lost forever.

There are several ways to eliminate reflected waves in an antenna system. You can use a trimer capacitor, rf transformer, or a shunt-feed arrangement such as an L-network. You'll find the trimmer capacitor is a method most often used in auto radios. When making an adjustment of the system, first tune the receiver to a point on the dial that is between stations on the AM band. You should hear only very weak noise. Then adjust the trimmer capacitor for maximum noise output. Incidentally, the antenna should be at normal receiving height during the entire procedure. If you can't get a good, sharp peak, check the transmission line and the antenna, because one or both may be defective.

Frequency Tracking Test

To begin, no alignment or tracking adjustments should be attempted on radio receivers unless it has been clearly established that the stages are actually mistuned (in other words, first check area signal strength, antenna system, etc.). When it has been definitely established that the set isn't tracking properly (you'll have weak reception and, possibly, interference), troubles are often encountered because a few important factors are overlooked or misunderstood. First, it's necessary to know just what a tracking alignment job is trying to accomplish. Simply stated, the whole idea is to maintain the proper frequency relationships in the receiver front-end circuits (mixer and local oscillator), which usually are designed to be simultaneously varied in frequency by ganged operations (see Figure 7-2). You'll find that the higher the frequencies you're working with, the tougher the tracking alignment job. For instance, FM radio (88 MHz to 108 MHz) requires a more painstaking procedure than does AM radio (535 kHz to 1605 kHz).

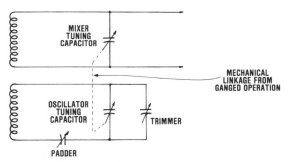

Figure 7-2:   Tracking adjustments in a receiver front end that uses a padder and trimmer for this purpose.

Another important point is to never forget the rules of working with high frequencies when you're working with rf circuits. These rules are:

1. All leads must be as short as possible and coaxial cable must be used when interconnecting test gear. If you include even one piece of unshielded wire a few inches long when connecting a scope or signal generator, it may cause your measurements to be in serious disagreement with the manufacturer's service notes.

2. Be sure that you connect a good ground between all test gear and the unit under test because a floating ground on the scope or signal generator can really upset your expected results.

A great number of receivers do not use a padder and trimmer such as the ones in Figure 7-2. Instead, the tracking is accomplished by varying several coils in the tuned circuits (see Figure 7-3). The tracking procedures for various sets differ but the preceding precautions apply to any tracking adjustments involving

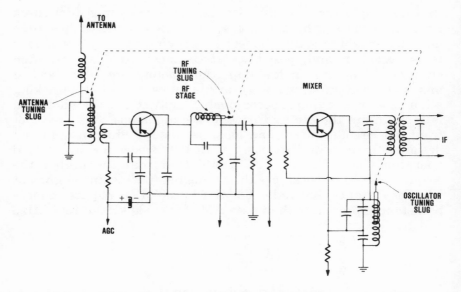

Figure 7-3:   Solid state radio receiver front end that uses variable
inductors for tuning. For proper tracking, the antenna, rf, and
oscillator coils must be properly aligned to produce the correct
frequency relationships.

a signal generator. It will be to your advantage to keep a few more
points in mind during the tracking adjustment procedure.

1. Use a suitable dummy antenna to connect the signal
generator to the receiver antenna. Although you should consult the
receiver's service data to determine the proper resistor values for a
dummy antenna, typically, component values and design will be
similar to one of the two shown in Figure 7-4.

2. As a general rule, it isn't wise to attempt a tracking
adjustment on a high-frequency receiver (for example, FM) if you
don't have quality test equipment. Figure 7-5 illustrates how a signal
generator and oscilloscope are connected for tracking the rf
mixer/oscillator stages. However, to repeat, if your scope has poor
sensitivity, you'll probably see no indication whatsoever when

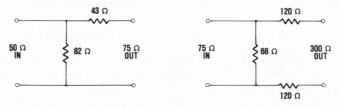

Figure 7-4:   Impedance matching pads, dummy antennas, or coupling
networks used to connect test equipment for receiver servicing

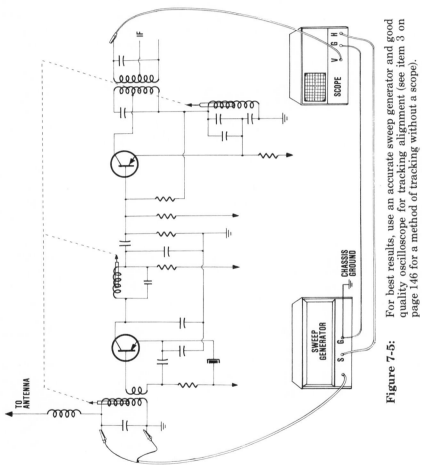

Figure 7-5: For best results, use an accurate sweep generator and good quality oscilloscope for tracking alignment (see item 3 on page 146 for a method of tracking without a scope).

connected to the receiver mixer output, as shown. Also, it's possible that your scope will overload the mixer. If you run into this trouble, try placing a 10 k ohm resistor in series with the scope probe and it should eliminate the problem.

3. You can do these adjustments using nothing but a variable-frequency signal generator and an AC voltmeter. Maybe not as good, but it'll work. To track the circuit, simply couple the signal generator to the receiver's antenna input, as shown in Figure 7-5. Then connect your AC voltmeter across the speaker terminals. This is your output indicator. Now, tune the receiver to the high end of the tuning dial ... say about 1615 kHz, if you're working on an AM broadcast receiver ... and set the volume control to maximum. Set your signal generator for a weak output signal. Now, it's simply a matter of following the procedures shown in Figure 7-2 or 7-3, depending on which of the two systems is used.

Troubleshooting the RF and Mixer Stages

When an "It won't play" receiver arrives at your workbench, one of the first steps is to apply power to the set and turn the volume control up to maximum. If you hear a noise output at the speaker, go straight for the local oscillator. There are two ways to check this stage:

1. Assuming the receiver DC supply voltages are the correct value, you can check the DC voltages at the emitter of the oscillator transistor. Connect the voltmeter between ground and the transistor emitter lead. Tune the receiver from the high end of the dial to the low end (actually, either way is okay). You should see a variation in voltage readings if the oscillator is operating properly. If you don't see the voltage change as you tune the set, it's an indication that the oscillator stage is nonoperative. It could be a bad transistor (particularly, if you read zero DC voltage). If it is, you may be able to parallel a new transistor alongside the one you think is bad. If you can and the set starts working, it will confirm your suspicions.

2. Another way to do the job is to substitute a properly tuned signal generator for the normal local oscillator. Connect the generator lead to the mixer through the oscillator coupling capacitor, or, sometimes, simply holding the generator test lead close to the oscillator circuit will work. If the receiver starts working, the oscillator is bad. Of course, this is assuming the generator is tuned to the oscillator's normal operating frequency for the frequency the dial is set to. Generally, the correct frequency can be found by using the formula LO = RF + IF, where LO is the local oscillator frequency, RF is the frequency the receiver is tuned to, and

IF is the receiver's intermediate frequency. In most cases, this will be 455 kHz or 262 kHz.

When you suspect the trouble is in the receiver front end, a few possible defects that may be found in these stages are:

1. A capacitor (the most troublesome component in solid state receivers).
2. A transistor is open, shorted, or leaky.
3. A cold solder joint.
4. A break in PC board wiring.
5. Incorrect replacement parts.

If you encounter a weak output, it's a bit more difficult to troubleshoot because this condition can be caused by almost any stage in the receiver. However, you can almost bet the problem is *not* a badly aligned front end unless someone became "screw driver happy" and didn't know what he was doing. Assuming there isn't an alignment problem, it's possible that you may run into a set that does not receive stations at the high frequency end of the tuning dial but works okay at the low end. Or, the set may work for a short while at the high end of the dial and then slowly get weaker and weaker until there is no signal at all. If you run into either of these situations, a prime suspect is the converter transistor.

On the other hand, you may have a fairly strong output signal but it's distorted. The stages in the front end may cause a distorted output signal and, in many cases, it's because the rf amplifier has started oscillating. A scope will quickly show you whether the rf amplifier is oscillating (look for an output signal with the antenna leads shorted). If the stage is operating in a regenerative mode, check the transistor. It may not be the correct replacement type. After you've made these checks, and if all circuits in the front end prove to be good, your next best bet is to start testing the IF stages.

Troubleshooting Solid State IF Amplifiers

To some extent, the intermediate frequency amplifiers are the most important stages in a receiver, at least so far as the selectivity and gain of the receiver are concerned. In the previous section, it was suggested that one of the first steps in troubleshooting a disabled receiver was to check the local oscillator. However, it isn't all that rare to find that the reason for a "dead" receiver is that one of the IF amplifier transistors has given up the ghost. Several other symptoms that may be caused by a defective IF stage are a weak output, distorted output, or a high noise level in the output.

Most IF amplifiers employ transistors connected in the

common-emitter configuration, although **IC's** are being used for IF strips in many modern receivers. A widely used technique that will give you an idea about which part of a transistor receiver is malfunctioning is to check the temperature of each transistor. The standard test is to try each transistor with your finger. However, if you're checking silicon power transistors, *watch it!* This is a good way to end up with a burned finger. A thermometer is a much safer way. This same trick can be used in any type set, transistor or **IC**, but checking with temperature will only locate the *area* of the fault. For example, if you find an IF transistor overheating, you'll still have to make measurements of voltage levels, current, and resistance, for accurate fault location. Just because the transistor (or any other component—resistor, coil, etc.) is overheating, it isn't necessarily a positive indication that the particular component is bad.

There are no two troubleshooting procedures with the same first step, but many servicemen start all weak signal output jobs by checking the IF stages. Inasmuch as front ends of receivers, as well as IF systems, differ to a considerable extent among the various manufacturers, no two receivers are aligned in exactly the same sequence, though general rules do apply. When undertaking *IF alignment* of any receiver, it's always best to refer to the detailed steps given in the service manual for that set. However, the procedures needed for receiver *IF troubleshooting* are not drasticaly different for most receivers. Take a CB receiver, for example. Like any receiver, the AM detector is a good—and usually convenient— place to connect your DC voltmeter as an output indicator. The reason is that, as the signal reaches the AM detector (this would be the ratio detector or limiter, in an FM set), the diode (or diodes) develops a DC component that is proportional to the IF signal variations and can be read with a DC voltmeter.

If you're working on a single-conversion receiver (many communications receivers use a double-conversion mixer) that has only one IF section, start by injecting an *accurate* IF signal from your signal generator into the base of the mixer transistor. As was explained before, connect the DC voltmeter to the AM detector output. Incidentally, never try to check a CB receiver IF section while the rig is operating single-sideband. This is because IF strips are adjusted for double-sideband operation and, therefore, all specs are given for this mode.

The three most common IF frequencies used in single-conversion CB rigs are 7.8 MHz, 10.7 MHz, and 12 MHz. Of course, you'll find that so many sets use the old, familiar IF frequency of 455 kHz that it appears to have become almost standard in receiver IF circuit designs. At any rate, when servicing, it's best to always check the schematic or service notes for the correct IF. *Remember*, in many cases, the signal generator you use *must be very accurate.* Be sure to check this out, particularly when working with communications receivers. When in doubt, use a frequency counter along with your signal source.

After you have completed the test hookup, adjust the generator output until you have a convenient DC voltmeter reading. *Do not* overdrive the rf and IF stages during the setup. Next, peak all the mixer and IF adjustments, if necessary. The highest DC voltmeter reading should coincide exactly with the IF design frequency. If it doesn't, the stage you're adjusting is at fault. The exception, and more than likely the trouble in CB receivers, is that the crystal has shifted frequency.

One trouble you may run into is oscillations in the IF stages. The symptom, when you're using a voltmeter connected to the detector, is a very high DC voltage reading. More than likely, you'll find that your voltage will shoot up when you're trying to peak a certain IF stage. When you run into this problem, it's generally better to replace the coil assembly or decoupling capacitor.

When troubleshooting IF stages, two precautions should be observed: use an isolating capacitor in series with the signal generator so the bias voltages will not be shunted by the test generator, and place a capacitor across the scope test probe between the hot lead and ground lead, to help reduce the pickup of spurious signals.

Another point worth repeating is that if the DC level, as read on the voltmeter connected across the detector, varies, the system is not overloaded. When the DC level remains constant with an increase in the generator output, you are overloading the circuit and

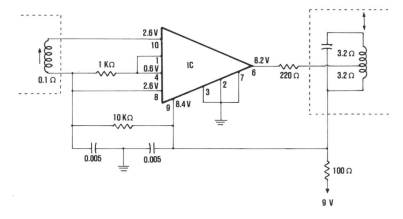

Figure 7-6:   Typical IC amplifier voltage and component values that can be checked with shop instruments

should decrease the generator output level below this point. Make this check before each IF stage adjustment.

Figure 7-6 shows voltage and component values for a typical **IC** amplifier. Since we have no access to the interior of an **IC**, troubleshooting an IF strip using **IC's** is more quickly and easily accomplished by measuring the voltages and checking the external components. Current measurements are also possible, but, unless you have a special in-circuit current measuring probe or use an indirect method (measure the voltage across a resistor and use $I = E/R$), you'll have to open the circuit to insert a milliammeter. Note that measurement of the resistance of some of the coils requires an ohmmeter with very low ohms reading capabilities. (See Chapter 2 for additional troubleshooting aids pertaining to receivers using **IC's**.)

Correcting Feedback Problems

Because oscillation in circuits such as IF amplifiers is caused by positive feedback from the output circuit to the input circuit, you'll find that many receivers use neutralization in their IF amplifier stages. Typically, there will be a neutralizing capacitor placed between the base and emitter (common emitter stage) of the IF amplifier transistor (see Figure 7-7).

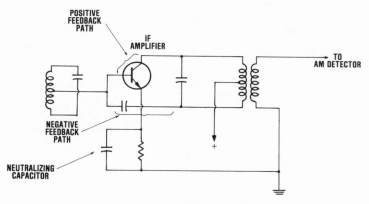

Figure 7-7:   IF circuit using capacitor neutralization

If you're having troubles with distortion in the output signal of a receiver, one possible cause is that the neutralizing capacitor is faulty. The purpose of this capacitor is to cancel the positive feedback that occurs from the collector of the transistor to the base (see Figure 7-7). To neutralize the stage, this capacitor must feed back a voltage that is 180° out of phase with the voltage fed back through the positive feedback path. If this capacitor is open or an incorrect value, it can't do its job and the results will be an unstable stage that may break into oscillation. This will, in turn, cause you to hear a distorted output signal from the receiver. *Note:* in severe cases of sustained oscillation in IF ampifiers, it is possible to completely kill the receiver output signal. In this case, the DC voltage that is measured at the receiver detector output will usually be much higher than the service notes show as a normal voltage.

Another tricky symptom, if you've never encountered it, is that the receiver output is distorted quite a bit when tuned to weak stations, but not too bad when tuned to a strong station. What may be happening here is that the AGC is producing more negative bias voltage on the strong stations (causing the IF stages to operate with a low gain) than on the weak stations. In this situation, the positive feedback (regeneration) is cut down on the strong stations by the AGC, resulting in very little, or no, oscillation. On the other hand, a weak station has weak AGC, strong regenerative feedback, and very noticeable distortion.

Servicing AFC Circuits

Many receiver systems (radio or TV) include an automatic frequency control (AFC) circuit. This circuit is almost always included in systems such as these because they operate at a comparatively high frequency and may have a tendency to drift. The function of an AFC circuit in any kind of receiver is to hold the oscillator frequency at exactly the correct frequency and phase. The correct frequency and phase for a TV receiver horizontal oscillator is exactly the same as the transmitter horizontal sync pulses. A radio receiver (FM or AM) must have its local oscillator frequency reset to some certain predetermined frequency each time it is tuned to a new station. With the utilization of solid state devices, circuits are being designed and employed that exert even more exacting frequency and phase control.

Phase-locked loops (PLL's) are a comparatively new class of monolithic circuit but they are based on frequency feedback technology that dates back over 40 years. The first widespread use of phase-lock, however, didn't really come about until the television era and development of complete single-chip **IC's**. Now, a single, packaged **IC** with a few external components will provide all the benefits of phase-locked operation, including independent center-frequency and bandwidth adjustment. Figure 7-8 is a block diagram of a phase-locked loop.

A phase-locked loop in a communications circuit denotes a local oscillator that is synchronized in phase and frequency with a

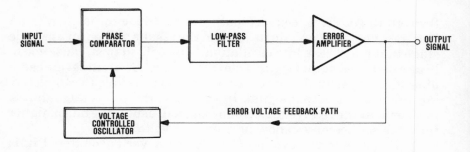

Figure 7-8:   Block diagram of phase-locked-loop

received signal, as we have stated. But, there are several ways this job is accomplished. As an illustration, in many FM tuners you'll find a 38 kHz oscillator, similar to the one shown in Figure 7-9, connected in a PLL configuration. You'll notice that the oscillator is synchronized by a 19 kHz pilot subcarrier.

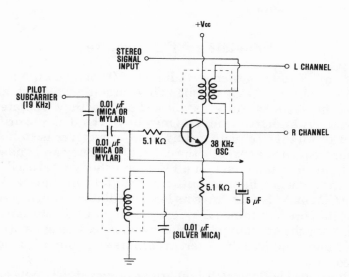

Figure 7-9:   An FM tuner 38 kHz oscillator circuit that will operate in a phase-locked-loop configuration

Phase-Locked Receiver Oscillator

A simplified block diagram of a phase-locked receiver oscillator and associated circuits is shown in Figure 7-10. Premixing is a form of frequency synthesis, therefore the entire circuit is often called a *frequency synthesizer*. It should be pointed out that this

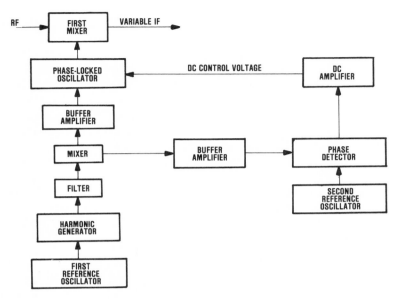

Figure 7-10:   Frequency synthesizer being used as a superheterodyne receiver local oscillator

conversion scheme is used in many different applications, test equipment, etc.

The phase-locked oscillator is controlled electronically by a DC voltage applied to a varactor diode. The output of the first reference oscillator (crystal controlled) is applied to a harmonic generator. A simple harmonic generator is no more than a tuned base, tuned collector radio frequency amplifier. The base tank circuit is tuned to the fundamental frequency—the crystal frequency, in this case—and the collector tank circuit is tuned to the second, third, or fourth harmonic. Notice, this will produce an output that is a set of discrete frequencies called a *comb* of frequencies. The appropriate frequency for the comb is selected by the filter following the harmonic generator. The filter is mechanically linked to the phase-locked oscillator. The frequency output of the filter is used as a reference to which the phase-locked oscillator is locked and is generally higher than the desired output of the phase-locked oscillator.

After the reference frequency is selected, it's combined with the output of the phase-locked oscillator in the mixer. The output of the mixer is the difference between the two frequencies. This difference frequency is applied to a buffer amplifier (or *isolation amplifier*, in that it isolates the phase detector from the mixer). Notice that the phase detector has two frequencies applied to it; the output of the mixer and another cyrstal controlled reference oscillator. *Note:* if the outputs of the mixer and second reference oscillator are the same frequency, there will be no DC output from the phase detector. If the DC control voltage is zero, there will be no change of frequency at the phase-locked oscillator. If the two frequencies are not exactly the same, the DC control voltage will not be zero. Therefore, the phase-locked oscillator will be tuned until the output is zero again. Of course, there is a limit to how far off frequency the circuit can be and still hold lock.

Frequently, there is a panel light that warns when the circuit is not locked. The procedure is to manually tune the phase-locked oscillator until the light goes off, indicating lock has been achieved. In this manner, it's possible to produce many crystal-controlled frequencies with only two crystals and no crystal switching is required, as it was a few years ago.

Troubleshooting and Repairing Modern PC Boards

If you've ever tried troubleshooting an electronic project where the wiring looks like a can of worms, you know this is, in itself, a major headache. It doesn't take anyone very long to find out that with printed circuit boards, troubleshoting is easier and, if the circuit requires it, you can make changes.

Sooner or later almost all of us end up trying to design a project, or add an improvement to a pre-printed circuit. Only if you are copying a preplanned circuit from a magazine, can you get by without sketching the layout for your board. In fact, this is the first step in constructing a new PC board or modifying an old one. When you start your sketch, be sure all components are included and placed in the best position for easy copper tape interconnections. Whatever you do, don't start work on the actual board until you are sure you have a satisfactory pattern worked out. If necessary, draw and redraw the circuit several times. What you want to have is the minimum number of crossovers and, above all, enough space for the components you intend to mount on the board. As for the supplies needed for the actual PC board itself, *Bishop Graphics*, 5388 Sterling Center Drive, P.O. Box 5007, Westlake Village, CA 91359, is, perhaps, the best known major supplier. They produce just about everything anyone who is interested in an easy way of creating or repairing PC boards would want.

When you use this *Quick-Circuit* system, all you do is remove an adhesive cover from a component mounting configuration, align drilled holes and press in place. For interconnecting components, they suggest the use of copper tape or jumpers. You can use insulating mylar tape between the copper tape, when making crossovers. Finally, insert your electronics components and solder the board. You now have a repaired or prototype PC board and, believe it or not, the whole process is almost as simple as it sounds. However, as in all electronics work, there are a few do's and don'ts when repairing or making PC boards. These are:

1. Don't try to solder a dirty surface on any PC board. Clean the surface to be soldered with either a rubber eraser or alcohol.
2. Above all, use only a *small amount of solder.*
3. If you don't already own one, it's almost a must to have an X-acto knife for working on PC boards. Numbers 5 and 7 are the blades most frequently used in this work.

Troubleshooting the Audio Section

The audio section of a low-cost solid state radio receiver includes all circuits from the output circuit of the detector to the speaker. Much information that you'll need for troubleshooting this section of a receiver is covered in Chapters 1 and 2. An analysis of the common trouble symptoms peculiar to radio receiver servicing is presented in this section.

1. **No sound out of the speaker:** Your best bet, under this circumstance, is to inject a signal into the output stage of the detector. If you do not hear sound at the speaker, more than likely, there is a defect in the audio section (assuming all DC voltages are present). Possible defective components that would cause a receiver not to produce an output due to trouble in the audio section are:

a. Defective capacitor (most common). See Figure 7-11.
b. Shorted transistor in the output amplifier (next most common).
c. Resistor open.
d. Bad diode in the detector.
e. Transistor in a stage preceding the output stage. Generally, it's best to check the output circuit transistors before testing any preceding transistor.
f. Speaker coil (possible but not likely).

2. **Sound coming from the speaker but no volume:** You'll find that, in most cases, the trouble will be the same components as those listed in item 1, except the capacitor, transistor, diode, etc., will be leaky, weak, or perhaps barely operating, due to low voltage. A good place to start your troubleshooting is to check the transistor-emitter bypass capacitor, if there is one. If such a capacitor opens, it causes the voltage developed across the emitter resistor to appear at

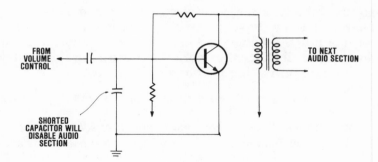

Figure 7-11: Example of an audio amplifier circuit showing the type capacitor connection to check when a receiver audio section is completely "dead"

the transistor input as a degenerative voltage (generally, you will not find an emitter bypass capacitor used except where high gain must be obtained from a single stage). Incidentally, if an emitter bypass capacitor does open you won't hear distortion on the output signal. In fact, it may improve the quality, but the output level will decrease. The easiest way to test the capacitor is to parallel the one you suspect with another capacitor of the same type. If the sound level increases, you've found the trouble. If not, check the transistor bias circuit, volume control, and tone control circuit. (Figure 7-12)

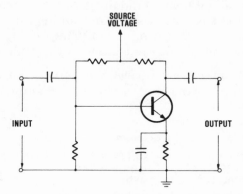

Figure 7-12: Audio amplifier circuit using an emitter bypass capacitor

3. **Hum and/or distortion:** The only time you'll have hum in a receiver output is when you're working on an AC set. Almost without question, you can go straight to the capacitors in the DC power supply, if you're working on a low-cost home receiver. Other causes of sound defects that may be due to troubles in the audio section are:

a. Transformer faulty.

b. Incorrect type transistor being used as a replacement.

c. Bad capacitor in negative feedback path.

d. A defective switch or control. Try an aerosol cleaner/ lubricator on switches and controls.

How to Cure AGC Troubles

Most receivers have some form of automatic gain control circuit. Many of the receivers now being manufctured employ IC's or a combination of discrete transistors and integrated circuits. In fact, when working on a solid state color TV receiver, it's common to find an AGC gate, IF AGC amplifier, and rf AGC delay stage all contained in a single **IC**. However, because we're talking about common broadcast and communications receivers, let's stick to their AGC circuits. Figure 7-13 shows a basic wiring diagram of such a circuit.

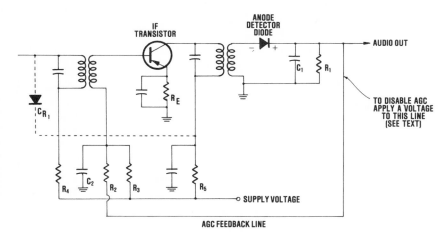

Figure 7-13: AGC circuit that could be employed in a common broadcast receiver to control an IF stage gain. This transistor requires a positive AGC bias voltage (see text)

Referring to Figure 7-13, you'll notice two diodes (C_{R1} and the detector diode). The detector diode and its load resistor, R_1, work together to produce an AGC bias voltage that varies with signal strength and is fed back to control the gain of the IF stage. You'll usually find an additional biasing network in the same receiver (C_{R1}). This, as shown, is called a *shunt diode*. The variable bias network applies a voltage that becomes more positive at the base of the IF transistor as the signal increases in strength—or less positive as the signal decreases in strength. The shunt diode (C_{R1}, shown with dashed line) has no affect on the IF circuit until the received signal (IF stage input) becomes quite strong; i.e., the two systems handle normal signal strengths and large signal amplitudes.

When troubleshooting an AGC circuit such as the one shown, pay particular attention to the values of R_4 and R_5, if C_{R1} is included. They set the reverse bias under no-signal condition for C_{R1}. The normal IF stage transistor bias is set by resistors R_1, R_2, and R_3.

Most AGC circuits in small transistor radios are straight-forward and almost identical. The service notes on many of these sets are not always available, but there are a few tricks that you can use to determine the voltages that should be present on particular test points. The AGC voltage has to be a DC voltage, either positive or negative, and this is determined by the type transistor used in the IF stage (or any other stage, for that matter). The key is that AGC is positive when applied to the base of a PNP transistor (see Figure 7-13), and negative when applied to the base of an NPN transistor. In either case, the AGC affects the emitter-to-base bias and adjusts according to an incoming signal and, when using a fairly strong input test signal, you should be able to see voltage changes with a voltmeter.

A quick way to determine if the AGC is positive or negative is to look at the polarity of the detector diode. If this diode is attached with its cathode connected to the AGC line, it's positive (see Figure 7-13). If the anode is connected to the AGC line, the AGC output will be negative and the IF transistor will probably be an NPN type.

When testing receivers, it's often necessary to disable the AGC circuit. The best way to do this is to connect a fixed DC voltage source, of opposite polarity and sufficient amplitude, to the AGC line (see Figure 7-13), to cancel the voltage normally produced on this line, as explained in the preceding paragraph.

Servicing the Squelch Circuit

The squelch circuit found on CB and other communications receivers is the opposite of the AGC circuits explained in the previous section in that it should totally "kill" the receiver output until a signal overrides some predetermined bias level set by the operator. Communications receivers usually include a delayed AVC (the terms *automatic gain control* and *automatic volume control* are usually interchangeable since they mean basically the same thing) because it is desirable to have maximum rf and IF gain available for weak signals. Incidentally, most of the time you'll find a manual

AVC cutoff switch that can be used to entirely disable the AVC for maximum receiver sensitivity. This is particularly desirable during times of emergency operation of the receiver, as when trying to receive a very weak distress call.

The AGC amplifier, and its associated circuits (found in better receivers) may be constructed using field-effect transistors and **OP AMP's**, or with the entire circuit contained within a single IC. In any case, when working with these circuits, follow the basic rules given in Chapters 1 and 2 for troubleshooting solid state circuits. It's important to note that the techniques given in Chapters 1 and 2 will uncover a defective **IC** as easily as any other component. There is very little difference between troubleshooting a squelch circuit and a simple AGC circuit. Just remember that the squelch circuit is supposed to keep the receiver inoperative (by applying a high bias) until the incoming signal is in excess of the squelch bias you set with the squelch control. Additionally, you need to know what the correct voltages in each section, stage, and circuit are when the receiver is operating normally. The best place to get this information is from the service notes, another receiver of the same type, or from a local sales/service center. Almost all manufacturers will have at least a sales representative somewhere in your vicinity. As the advertising blurb says, "Let your fingers do the walking." Use the yellow pages in the phone book. Many times a manufacturer's representative can tell you where to get a service note or schematic for the piece of equipment you're having trouble with.

CHAPTER 8

How to Test and Troubleshoot
Present Day Digital Circuits

This chapter provides practical guidance for troubleshooting today's digital equipment. You'll find there are many simple procedures for servicing digital integrated circuits as well as a few sophisticated ones but, in either case, the emphasis is not only on *how* to troubleshoot these circuits but also on understanding *what* you're doing.

When you're troubleshooting **hardware** (a term used to describe the physical or hard equipment that makes up computer systems), the chief requirement is to learn to think digital—in other words, visualize a switching problem from two points of view; as a logic structure and as a circuit. The following pages will cover both these approaches to digital servicing and include many specific ways to troubleshoot and test discrete systems such as **IC's**. Also, you'll find that much practical information on electronic instruments used for servicing digital integrated circuits, is included.

Another area many so-called "old timers" are still not comfortable with is the computer-oriented language (sometimes called *computereze*) used to describe logic structure and digital circuits. Therefore, you'll find that all technical terms are thoroughly explained throughout the chapter.

Peculiarities of Modern Digital Equipment Servicing

Isolating a fault in a digital circuit can be both difficult and time consuming if you don't know how to go about it. How do you service these units? The answer doesn't have to be "with great difficulty," if you understand logic. You may think that you know nothing about logic but don't you believe it! You've been using logical troubleshooting procedures ever since you first started repairing electronic circuits. Fortunately, you'll find that a whole lot of the circuitry is divided into separate logic blocks, and you can still use old-fashioned logical troubleshooting thinking to pinpoint the function or stage that is not doing its job properly. To replace it, you simply change the bad component (such as an **IC** or module). Take a video game, for example. If one player has no control functions but the other does, you immediately know which section isn't working.

You'll find that TV tuners are the same. If a set tunes to the correct channel but there is no digital read-out, simple logic tells you where to start checking. You're probably saying "Oh sure, so far it's easy, but what about digital test equipment and troubleshooting individual **IC's** and the logic family associated with them?" Okay, good question. First, to troubleshoot digital **IC's**, it's handy to have the manufacturer's service data. This data includes pinouts and functional truth tables. Also, it's a great help to have a good understanding of the particular **IC's** intput/output characteristics (we'll examine these points in the following pages).

How about special equipment? Can you use your old gear? Well, two pieces you'll certainly use are the *multimeter* and *oscilloscope*. As always, use your multimeter to check DC voltages being fed to the faulty circuit (you can also troubleshoot logic levels, but we will get to that soon). Just be sure to remember that some circuits use a dual-polarity DC power supply and some use a single-polarity supply. See Figure 8-1 for a basic schematic of each type.

Because an **IC** must be tested as a complete circuit, the scope is one of the most useful pieces of test gear you have. However, for the best results, the scope should have dual-trace with a triggered sweep. In fact, you'll find that many older scopes (except for limited circuits—such as checking clock oscillators for operation, etc.) just can't do the job.

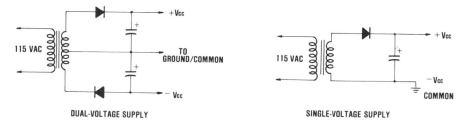

Figure 8-1:   Basic diagrams of a dual-polarity and single-polarity DC power suppy

Since most of us come into contact only with **TTL, CMOS** and **MOS** devices, a logic probe for **TTL/CMOS** digital circuits will usually take care of the oscilloscope problem. One thing for sure, a logic probe is a lot less expensive than a new scope. A probe generally indicates both *logic high* and *logic low* levels. It also indicates the polarity and *presence* of signal pulses.

When checking logic levels, absolute amplitudes are unimportant. A digital signal has three states and only two of these are used to transmit data; logic high (1) and logic low (0). In case you're wondering how the third state is identified, it may be designated as undefined, i.e., somewhere between logic high and logic low. Or, more common, the third state represents a high impedance, with no signal output. The term "state" refers to the voltage on the output or input of a device. For example, for transistor-transistor logic (**TTL**), when the amplitude of the signal is 0.8 ±0.15 V, it's at logic 0 level. When the amplitude of the signal is 2.1 ±0.25V, it's at logic 1 level. Anything in between these amplitudes is sometimes designated as undefined. This should give you an idea about the job of the logic probe, which is to show intermediate or "bad" logic levels. Also, the logic probe can detect pulses of short duration (typically, as short as 10 nano sec). These pulses can be a problem when using inexpensive scopes.

Four other special digital troubleshooting instruments are the logic clip, logic comparator, logic analyzer, and logic pulser. The logic clip is normally used for checking during quiescent (no signal being applied) conditions, or with low-frequency signals. The logic comparator is designed to compare one **IC** to another (some of these comparators come in the form of a clip). Of the four, the logic analyzer probably is the best. For instance, this instrument can be programmed for the examination of streams of sequential data. It is particularly good for checking calculators and other **ROM** controlled systems. The last instrument listed, the logic pulser, is a pulse generator. The output pulse is used to cause an **IC** to change states during troubleshooting. Some servicemen use standard pulse generators to do the same job but, in most cases, this type instrument is difficult and may slow the day's production quite a bit. Generally speaking, a multimeter, scope, pulse generator, and logic clip or digital probe will be sufficient test gear for the jobs encountered in an average service shop. Other than these, you'll probably have to have a set of **IC** test clips and jumper leads.

How Truth Tables Help During Troubleshooting

Although logic analyzers will display digital circuit truth tables on a CRT screen, it's also possible to produce a truth table by: 1) working one up yourself; or, 2) consulting the manufacturer's service notes for the circuit under consideration. You'll find that truth tables are a practical aid in analyzing circuits. For instance, when you use an oscilloscope all signals must be checked to see if they are at a valid logic level. A truth table is a statement of the states of all inputs and outputs of a given circuit under a specific condition. You'll find that it can be an extremely helpful tool, especially when working with a complex system. Remember, *you must look at logic levels, not waveforms* when troubleshooting digital equipment.

Figure 8-2 is the symbol and truth table for an *Exclusive-OR gate*, one of the more frequently used logic circuits. Referring to the truth table (Figure 8-2B), you'll see that when both input signals (A and B) are at the same logic level (either 0, 0 or 1, 1), a logic 0 appears at the output. Or, when the inputs (A and B) are at different logic levels (0, 1 or 1, 0), the output is logic 1.

A	B	OUTPUT
0	0	0
0	1	1
1	0	1
1	1	0

(A) SYMBOL

(B) TRUTH TABLE

Figure 8-2: Symbol and truth table for an exclusive-OR gate

Generally, you'll find more than one exclusive-OR gate in a single **IC**. For example, Signetics Quad 2-input exclusive-OR gate 7486 is a **TTL IC**, providing the outputs shown in Figure 8-3(b), with the inputs shown in Figure 8-3 (B) under A and B. Figure 8-3 (A) is equally important during troubleshooting because, as always, pin identification is important for almost any **IC** service job.

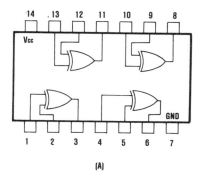

INPUTS		OUTPUTS
A	B	Y
0	0	0
0	1	1
1	0	1
1	1	0

(A)

(B)

Figure 8-3: Pin identification and truth table for a 7486 Quad 2-input exclusive-OR gate

Exclusive-OR gates are used for comparing one or two-bit numbers, but four-bit magnitude comparators are, by far, the best way to go in more advanced applications. Therefore, if you're an experimenter or serviceman, more than likely you'll be working with one or more of these **IC's** sooner or later. A functional block diagram of Signetic's 4-bit magnitude comparator 7485 is shown in Figure 8-4 (courtesy of Signetics Corp., Menlo Park, California).

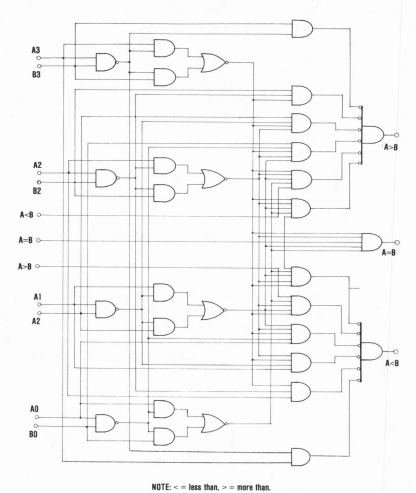

NOTE: < = less than, > = more than.

Figure 8-4:   Functional block diagram of a 7485 comparator that could be used as a troubleshooting aid

Referring to Figure 8-4, you'll see that the 7485 provides three outputs; (A > B), (A = B), and (A<B). Three fully decoded decisions

about two input nibbles (two 4-bit words, A, B) are made and are externally available at the three outputs. In other words, the device will tell you which of the two nibbles is the larger, or if they are equal.

Figure 8-5 is a pin diagram for the 7485. Notice the easy layout for troubleshooting, designing, and so forth. Also, the cascading inputs (A<B, A=B, and A>B) may be used to cascade inputs of two or more of the devices so that it's possible to compare more than the two 4-bit words (for instance, words having eight or more bits). Any experimenter or hobbyist will find the cascading inputs useful when working with this chip. This is particularly interesting when you consider that most personal computers work with words having eight or more bits.

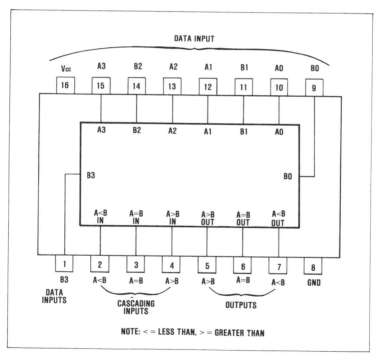

Figure 8-5: Pin configuration of a 7485 4-bit magnitude comparator

The truth table for a 7485 is shown in Figure 8-6. You'll notice that the outputs are designated as H and L. These two letters are used to indicate either a logic high (H) or logic low (L) output. In this unit, the logic high output voltage minimum is 2.4 V. The logic low output voltage maximum is 0.4 V.

To use the example truth table (Figure 8-6) for making the various parameter measurements, you'll need a pulse generator. The

COMPARING INPUTS				CASCADING INPUTS			OUTPUTS		
A3,B3	A2,B2	A1,B1	A0,B0	A>B	A<B	A=B	A>B	A<B	A=B
A3>B3	X	X	X	X	X	X	H	L	L
A3<B3	X	X	X	X	X	X	L	H	L
A3=B3	A2>B2	X	X	X	X	X	H	L	L
A3=B3	A2<B2	X	X	X	X	X	L	H	L
A3=B3	A2=B2	A1>B1	X	X	X	X	H	L	L
A3=B3	A2=B2	A1<B1	X	X	X	X	L	H	L
A3=B3	A2=B2	A1=B1	A0>B0	X	X	X	H	L	L
A3=B3	A2=B2	A1=B1	A0<B0	X	X	X	L	H	L
A3=B3	A2=B2	A1=B1	A0=B0	H	L	L	H	L	L
A3=B3	,A2=B2	A1=B1	A0=B0	L	H	L	L	H	L
A3=B3	A2=B2	A1=B1	A0=B0	L	L	H	L	L	H

NOTE: H = HIGH LEVEL, L = LOW LEVEL, X = IRRELEVANT

Figure 8-6: Truth table for a 7485.

pulse generator (in this case) must have a 50-ohm output and be set to operate at 1 MHz, with a duty cycle equal to 50%. *Duty cycle* is usually calculated by multiplying pulses-per-second times pulse width. Signetics suggests that you use the load circuit shown in Figure 8-7 to make the various measurements.

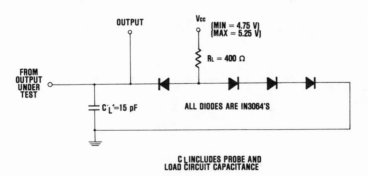

Figure 8-7: Load circuit for testing a 7485

Troubleshooting Gate Circuits

The three elements of **IC** logic are AND gates, OR gates, and INVERTERS. They are shown symbolically in Figure 8-8. *Note:* some manufacturers may use other symbols, but these are the most widely used today.

The INVERT symbol, as a circle, can be added to the AND and OR symbols to form NAND (AND inverted) and NOR (OR inverted), as shown in Figure 8-9.

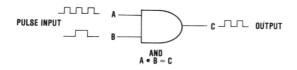

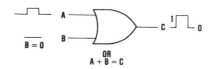

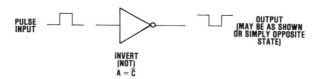

Figure 8-8: Three symbols of the basic elements of IC logic. Read the dot (●) as AND, the plus (+) as OR, to determine input-output signal results.

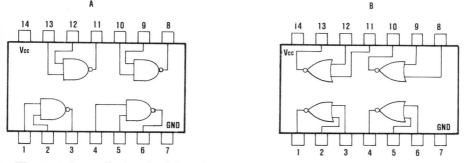

Figure 8-9: Section A of this illustration is an example of a quadruple 2-input positive NAND gate by Signetics. Section B is a quadruple 2-input positive NOR gate by the same company (7400 and 7402 respectively)

Referring to a single 2-input NAND gate, logic 1 input voltage is required at both input terminals to insure logic 0 level at the output. On the other hand, logic 0 input voltage is required at either input terminal to insure a logic 1 level at output. Next, the NOR gate. The 2-input NOR gate produces an output when both of the inputs are logic 0 level. If either of the inputs is 1, the output is always a logic

0 level. As may already be apparent, a truth table is a fast and simple way to "see" what you should find on the output of a 2-input gate during troubleshooting. The operation of a 2-input NOR gate is represented in the truth table shown in Figure 8-10.

INPUT		OUTPUT
B	C	$A = \overline{B + C}$
0	0	1
0	1	0
1	0	0
1	1	0

Figure 8-10:   Truth table for a 2-input NOR gate. A powerful tool, especially when troubleshooting a complex system of gates.

You'll encounter both negative and positive logic. To see what happens here, let's take the truth tables for a positive and negative 2-input AND gate and compare them (see Figure 8-11). The main thing to notice is that the truth table for the negative logic AND gate is exactly opposite to the truth table for the positive logic AND gate. Also, when you're checking logic circuits, it's helpful to remember that the negative logic AND gate acts as a positive logic OR gate (see Figure 8-11). In fact, an AND gate can provide the OR function and an OR gate can provide the AND function. It just depends on whether a positive or negative logic level is used.

POSITIVE LOGIC 2-INPUT
AND GATE
TRUTH TABLE

INPUT		OUTPUT
B	C	A
0	0	0
0	1	0
1	0	0
1	1	1

NEGATIVE LOGIC AND GATE
TRUTH TABLE

INPUT		OUTPUT
B	C	A
1	1	1
0	1	1
1	0	1
0	0	0

Figure 8-11:   Truth tables for positive and negative logic that could be applied to a 2-input AND gate

Normally, you'll find that *positive* logic means 1 is high and 0 is low. However, when *negative* logic is used, logic level 1 is low and 0 is high. The significant result is as explained in the preceding paragraph. Furthermore, an AND gate for positive logic becomes a NOR gate for negative logic and an OR gate for positive logic becomes a NAND gate for negative logic. All this explains why it isn't necessary to have both AND and OR, or NAND and NOR gates, but it also explains why truth tables are so helpful when troubleshooting or designing digital circuits using IC's.

How to Service Equipment Containing
Adders and Subtracters

You'll find digital adder **IC's** are basic to logic computers. There are several types; half-adders, full-adders, serial adders, and parallel adders, just to name a few. The half-adder is formed by using an exclusive-OR gate and AND gate. The output of this arrangement is the sum output A + B and the carry output AB. To put it another way, if A and B are both 0, the sum is 0; if either input (A or B) is 1, the sum is 1; but if A and B are both 1, their sum is 0 with a carry 1. To see it still another way, study the truth table in Figure 8-12. This illustration also shows the logic circuit. You'll notice that the logic circuit and truth table show only two inputs (A and B). This is important because it's one way to distinguish between a full-adder and half-adder. The full-adder will accept a third input, a "carry" input, as well as the two numbers to be added (A and B inputs).

CIRCUIT

TRUTH TABLE

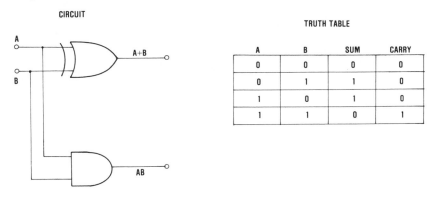

A	B	SUM	CARRY
0	0	0	0
0	1	1	0
1	0	1	0
1	1	0	1

Figure 8-12:   Logic circuit and truth table for a half-adder

Basically, the full-adder is formed by combining two half-adders. The gated full-adder of Figure 8-13 is a Signetics 7480 single-bit, binary adder with gated complementary inputs, complementary sum (Σ and $\bar{\Sigma}$) outputs and inverted carry output. The pin configuration and truth table for this device are shown in Figure 8-13.

Common trouble symptoms caused by defects in binary adders and subtracters are:

1. An incorrect readout.
2. No readout.
3. Readout unstable.

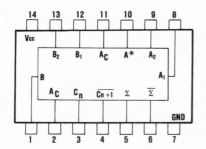

LOGIC					
Cn	B	A	$\overline{Cn+1}$	$\overline{\Sigma}$	Σ
0	0	0	1	1	0
0	0	1	1	0	1
0	1	0	1	0	1
0	1	1	0	1	0
1	0	0	1	0	1
1	0	1	0	1	0
1	1	0	0	1	0
1	1	1	0	0	1

Figure 8-13:   Pin configuration and truth table for a gated full-adder

When servicing digital equipment, you are concerned only with logic input/output states, either low or high. As we have said, remember, *logic low* stays near zero volts—in almost every case, 0.2 volts or so. *Logic high* will be anywhere from about 4 volts to as high as 16 volts, depending on what you're servicing. To troubleshoot **IC** logic systems, you only have to know a couple of things. Which output state should the circuit under test produce, and is the circuit producing that output? Now, once you know the output is supposed to be a logic low or logic high, you're in, because a logic probe or a plain old-fashioned voltmeter will usually verify these states.

The procedure using the voltmeter is easy. Inject a logic low and then a logic high—check the input and output with the voltmeter. If you're working with a noninverting stage, the voltmeter should read the same logic level at both input and output. An inverting stage will read exactly the opposite on the output, in reference to the input.

When incorrect readings occur, one way to start trouble-shooting is to grab a logic probe. The first step is to isolate the malfunction to as few **IC's** as possible. If you're working on a complex system, most of your time will be spent looking for the problem **IC**, not repairing it. To use the probe, trace logic levels and pulses through the various **IC's** to determine whether the point you're checking is high logic, low logic, improper level, open circuit, or pulsing. Once you've found the bad circuits, check with one pulse at a time, while referring to the truth table or schematic for the particular **IC** or equipment under test. You'll probably need a logic pulser with single-shot capabilities for this purpose.

If you have no read-out, check supply voltages, clock generator for clock pulses, and for **IC's** not properly mounted in their sockets. Erratic operation was also mentioned in the beginning of this section. If this is the case, it's very possible that your troubles are due to a faulty PC power supply. Check to see that the DC voltages are correct and stable. Of course, a bad solder joint is always a possibility, especially if the equipment has recently been repaired.

Troubleshooting BCD Counters

You will encounter many different applications of binary-coded decimal (BCD) counters when servicing circuits that incorporate digital logic. Unfortunately, most of us were taught that one has to learn *binary arithmetic* before we can *troubleshoot* circuits such as these. However, if you approach a working technician and ask him to explain digital logic using Boolean Algebra, in most cases he'll tell you "Don't need it." The truth is, you *don't* need binary arithmetic to troubleshoot binary counters, computers, video cassette recorders, or any other type of equipment, circuits, or stages that incorporate digital logic. But, it will expand your knowledge and is a help when designing circuits. See Chapter 10 for this information.

Let's take a look at a binary counter by Signetics (the 82S91) for an example. This is the ultra-high-speed version of the 8291. We'll see how to test it and understand basically what the internal circuits are, without using binary arithmetic. To begin, the package (top view) and logic diagram for the **IC** are shown in Figure 8-14. Counters and frequency dividers are basically the same thing but with different uses of the resulting outputs. In either case, you'll find they all use J-K flip-flop circuits.

Notice that there are four J-K flip-flop circuits in this **IC**, which is typical. Generally, the state of a given flip-flop changes when its clock input goes from 1 to 0. If we place four J-K flip-flops in cascade, it's possible to make a simple binary counter. When the first input pulse (from 1 to 0) triggers the first flip-flop, its output is 1. When the second input pulse goes to 0, the first flip-flop goes to 0 and the second flip-flop output goes to 1, and so on.

It's also possible to use such a circuit as a frequency divider (counter) with division factors of 2, 4, or 8. As an example, referring to Figure 8-14, this **IC** allows separate use of the first and last three stages. In the first stage, the input count frequency presented to clock pulse 2 appears at outputs BO = ÷ 2, CO = ÷ 4, and DO = ÷ 8 simultaneously. Similar action can be seen by setting up the test circuit recommended by Signetics and shown in Figure 8-15.

Counters can divide frequency by almost any factor. In fact, using the proper hookup, these **IC's** can divide by 2, 4, 8, 16, and so on. But, this generally requires a few more components. If the flip-flops are connected to a NAND gate and the output of the gate is connected to reset inputs (see Figure 8-14), the counter will count or divide in larger steps. Troubleshooting of counter circuitry or **IC's** is

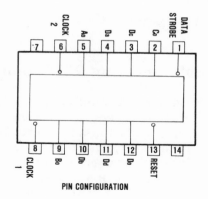

PIN CONFIGURATION

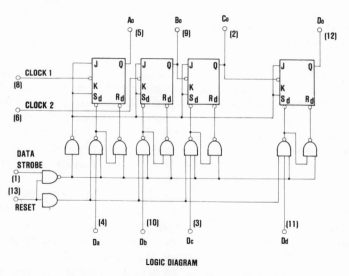

LOGIC DIAGRAM

Figure 8-14:   Package and logic diagram for a binary counter

simply a matter of checking logic levels. You can use your voltmeter and make simple *static* voltage measurements. However, whichever way you troubleshoot the circuit, you will need three things. These are:

1.   A schematic showing what should be read at each output.

2.   A way to check the input/output logic ... either a voltmeter, scope, or digital probe, etc.

3.   Remember that logic low readings usually will be in the 0.2 V vicinity, whereas logic high will most frequently be around 4 V or so. But watch it! As was said before, sometimes logic high can be much higher ... like 16 volts.

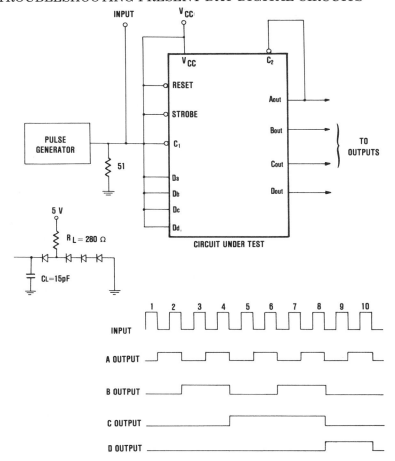

Figure 8-15: Test setup for checking the toggle rate of a binary counter. *Note:* the term *toggle* is used with reference to flip-flops connected as a counter, in which the process of changing state is called *toggling*

Servicing the Decoder/Driver Section

Decoder/drivers are used to switch the various segments in a display device (for example, liquid crystal or light-emitting diodes) off or on, in order to display numbers as well as some selected signs and letters. Decoder/driver **IC's** can be tested with a logic probe and logic pulser. However, let's first examine a seven-segment display arrangement that uses two **LED's** per segment, to get some idea of what a decoder/display driver may be driving (there are several other types of display devices—neon bulbs, Nixie® tubes, cold-cathode display tubes, and the RCA Numitron tube).

Figure 8-16 (A) shows a typical segment arrangement used by many manufacturers, and Figure 8-16 (B) shows a schematic diagram of a two-**LED** per segment (each segment is labeled A, B, C, etc.).

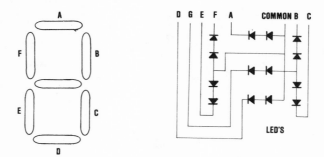

Figure 8-16:   Seven-segment display arrangements. (A) is a typical segment identification drawing. (B) is a two-LED per segment schematic diagram

Next, let's use Signetics' seven-segment decoder/display drivers #8T06 as an example display driver. The 8T06 accepts a 4-bit binary code and decodes all possible inputs as decimals 0-9, or as selected signs and letters. The pin configuration and package for this **IC** are shown in Figure 8-17. Other than inputs/outputs (a, b, etc.), V_{CC}, and ground, you'll notice there are some auxiliary inputs provided to increase the versatility of the unit. The ripple blanking input (RBI) and ripple blanking output (RBO) may be used for automatic leading and/or trailing edge zero suppression. The RBO output also acts as an over-riding blanking input (B1) that may be used for intensity modulation of the display. One other, the LT input, is provided to check the display by activating all outputs independent of the input code.

The fastest, simplest way to check for a defective **IC** is to interchange it with a duplicate **IC**. However, in some cases, it's best to have a truth table for the **IC**. A truth table for the 8T06 is shown in Figure 8-18. (*Note: always* remove power before removing and/or replacing an IC). This truth table shows a correct display character with the various inputs, and what the **IC** output states should be.

When troubleshooting, you may see other than these ideal situations. For example, you might see a zero read-out on all display devices. In this case, don't bother checking the DC supply because you are reading all zeros on the viewing screen, which means you do have power. Your first step would be to check for input pulses to the **IC**. As always, look for shorts, opens, and, possibly, an intermittent **IC**, especially if the trouble comes and goes.

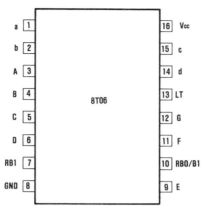

Figure 8-17: Pin configuration and package of a seven-segment decoder/ display driver 8T06

TRUTH TABLE

INPUTS						OUTPUTS							DISPLAY CHARACTER
INPUT CODE				LAMP TEST	RB1	B1/RBO	OUTPUT STATE						
d	c	b	a	LT	RB1	NOTE	A	B	C	D	E	F	G	
X	X	X	X	0	X		1	1	1	1	1	1	1	𝐵
X	X	X	X	1	X	0	0	0	0	0	0	0	0	BLK
0	0	0	0	1	0	0 (NOTE 1 & 2)	0	0	0	0	0	0	0	BLK
0	0	0	0	1	1	0	1	1	1	1	1	1	0	0
0	0	0	1	1	X	(NOTE 2)	0	1	1	0	0	0	0	1
0	0	1	0	1	X	1	1	1	0	1	1	0	1	2
0	0	1	1	1	X	1	1	1	1	1	0	0	1	3
0	1	0	0	1	X	1	0	1	1	0	0	1	1	4
0	1	0	1	1	X	1	1	0	1	1	0	1	1	5
0	1	1	0	1	X	1	0	0	1	1	1	1	1	6
0	1	1	1	1	X	1	1	1	1	0	0	0	0	7
1	0	0	0	1	X	1	1	1	1	1	1	1	1	8
1	0	0	1	1	X	1	1	1	1	0	0	1	1	9
1	0	1	0	1	X	1	0	0	0	0	0	0	1	-
1	0	1	1	1	X	1	0	0	0	0	0	0	0	BLK
1	1	0	0	1	X	1	1	1	1	0	1	1	1	𝐴
1	1	0	1	1	X	1	0	0	1	0	0	0	0	'
1	1	1	0	1	X	1	0	0	0	1	1	1	0	𝐿
1	1	1	1	1	X	1	0	0	0	0	0	0	0	BLK

X = DON'T CARE. EITHER "1" OR "0"
B1/RBO IS AN INTERNALLY WIRED OR OUTPUT.
NOTE:
1. B1/RBO USED AS INPUT.
2. B1/RBO SHOULD NOT BE FORCED HIGH WHEN a, b, c, d, RB1 TERMINALS ARE LOW, OR DAMAGE MAY OCCUR TO UNIT.

Figure 8-18: A truth table for an 8T06 seven-segment decoder/display driver

Sometimes a logic probe and logic pulser will show the circuit to be good and, when you place the equipment back in operation, it will not perform in a normal manner. In this case, your best bet is to check with a fairly good triggered oscilloscope for low pulse amplitude or distorted pulses.

Servicing Binary Shift Registers

By connecting the four JK flip-flops (shown in Figure 8-14) as shown in Figure 8-19, it's possible to end up with a so-called *shift register*. With this "hookup," the state of any given flip-flop is passed on to the next following flip-flop upon each clock pulse (the pulse generator in this arrangement is generally called a *clock*). In other words, the first flip-flop receives the input state on a given clock pulse and a pulse is passed along to each following flip-flop upon each succeeding clock pulse.

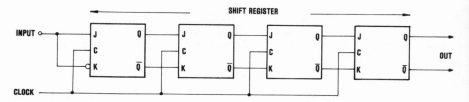

Figure 8-19:   Cascading JK flip-flops to form a basic shift register

The simple serial counter without feedback is known as a *ripple counter*. In this mode of operation, the clock may be derived from a pervious stage output. Another capability of some shift register **IC's** is *parallel processing* (having more than one input/output). Figure 8-20 shows the pin configuration of a 4-bit shift register (Signetics 82S70), having both serial and parallel data entry capability and parallel outputs.

This shift register has three operation modes; 1) synchronous parallel load, 2) right shift and, 3) hold (do nothing). However, by using two of the **IC's**, it's possible to end up with a left-right shift register plus other applications. A binary left-shift register will shift any binary number to the left after it has been entered, and do an "end-around" shift from the last flip-flop in the chain to the first. The easiest way to understand a shift register is to refer to its function table (see Figure 8-21). These registers' mode of operation depends on the logic levels on the shift and load inputs, as shown by the function table.

With both the load and shift inputs in the logical 0 state, the registers are in the hold mode and the system clock may be left free running without changing the contents of the register. This means that four bits of information can be stepped into a shift register and held there until they can be used. When the time comes for their use, the four bits are stepped out of the shift register. This is a common use of the device; temporary storage or memory. Parallel loading or serial shift is accomplished on the falling edge of the clock, when the

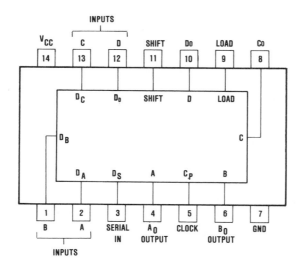

Figure 8-20: Pin configuration of a 4-bit shift register having both serial and parallel data entry capabilities

FUNCTION TABLE

| CONTROL INPUTS | | OPERATING MODE |
LOAD	SHIFT	
0	0	HOLD
1	0	PARALLEL LOAD
X	1	SHIFT RIGHT $(Q_A \rightarrow Q_D)$

Figure 8-21: Function table for a 4-bit shift register having three modes of operation (see text)

shift and load inputs are conditioned as shown in the function table (Figure 8-21).

Note that a clock pulse cannot trigger a flip-flop except when suitable J and K inputs are present. If you have an incorrect output (a 1 indication when a 0 should be indicated), it's very possible that the clock pulse has insufficient amplitude or is incorrect in some other way (both high and low logic levels being received by the fli-flops at the same time, or some other form of distortion). Check with a logic probe, clip, or scope, etc. As was explained before, one of the easiest ways to check **IC's** is to change the suspected **IC** with a known-good **IC** taken from the same piece of equipment you're servicing. *Note: always turn power off before removing and/or replacing any **IC**.* Also, it's possible that the input or output of the **IC** is shorted to ground or V_{CC}, or the input or output circuit is open. And, look for possible shorts between the **IC** pins. Any one of these troubles can cause occasional circuit malfunctions, or even a complete failure due to excessive heat buildup within the **IC** (especially when pins are shorted together).

Troubleshooting Modern Display Sections

Previous sections of this chapter briefly mentioned that there are several types of display devices used for read-out counters in digital equipment. For example, incandescent lamps, Nixie® tubes, Numitrons, Panaplex plates, **LED's**, liquid crystal, and a few more. Of course, this means that you'll encounter several different **IC's** designed to work in each system, especially when you're servicing older equipment. The good news is that there are only about five kinds of failure you're likely to encounter when servicing **IC's** in any section in digital equipment—including display. These are:

1. Failure of the **IC** (internal logic).
2. One or more pins shorted to ground.
3. Two or more pins shorted together.
4. One or more pins shorted to V_{CC}.
5. Open circuit in either input or output.

Some common symptoms you'll see on the read-out, due to troubles in the display or other read-out circuits are:

1. A completely blank screen (always off), or the screen shows no change (glows continuously).
2. Read-out incorrect.
3. Equipment screen will not start measurements, counting, or other read-outs, at zero.
4. Occasional circuit malfunctions causing intermittent readings on a screen, typewriter, or other read-out device.

Any one of these symptoms usually results from one of the five basic **IC** circuit failures listed in the beginning of this section. For example, a common cause for Nixie tubes glowing a steady zero read-out is a short in the equipment's counter inputs. In any case, check the driver circuit with a scope and/or logic probe, and logic pulser.

How to Use Scopes, Pulsers, Logic Probes, and Logic Clips to Service Digital Circuits

Testing digital integrated circuits is important to determine whether or not a particular **IC** is up to the manufacturer's specifications. However, users of only a few **IC's** can't very well afford sophisticated test gear. In fact, *really good* logic clips, logic analyzers, and, even more expensive, logic state analyzers can (and usually do) run into quite a bit of money. How can a small shop get around this problem? Well, the best way I know is to use the scope and other equipment you have and purchase no additional equipment except when absolutely necessary. As was mentioned in the beginning of this chapter, an oscilloscope can be used in much digital troubleshooting, especially if you have a dual-trace unit with triggered sweep. If you don't have a dual-trace with a triggered sweep scope and attempt to troubleshoot a logic circuit with an older, inexpensive one, it's important you realize that many common logic circuits operate at speeds that are high compared to the bandwidth of these scopes. You'll find it's impossible to make both time and frequency measurements. Furthermore, the vertical axis isn't generally calibrated, thereby preventing threshold measurements.

Regardless of these limitations, you can use any scope in some circuits when troubleshooting digital equipment such as TV's, electronic instruments, etc. For example, if the signal you're tracing is a moderately high frequency on only one line, by all means use a scope. If you're attempting to observe pulses in logic circuits, a × 10 (low capacitance) probe should be used. Keep the probe ground clip as close to the point of measurement as practical and *keep your ground lead short*. Also, *don't* use coax for direct connections— always use a probe. Coax cable may introduce excessive capacitance into the test setup which, in turn, may temporarily disable circuits such as flip-flops. Furthermore, with or without a dual trace scope, everytime you move the × 10 probe from one test point to another, *readjust the probe*. Compensated probes can introduce considerable error when working with high frequencies, if not properly adjusted.

Another piece of equipment you'll need, if you don't already have one, is a logic clip (also called a *logic monitor*) and/or logic probe. Both of these instruments overcome the oscilloscope problem. You'll find that these two are very good for observing static or low frequency data. Many of these units have a pulse stretcher and/or memory, which helps when you're trying to observe short pulses.

Logic clips examine several circuits at the same instant. The clip is simply several single probes all contained in one instrument package and functioning simultaneously. The clip is piggybacked onto an **IC**. The power to operate the instrument is drawn from the circuit under test and **LED's** indicate all logic levels (typically, **LED** on = logic 1 and **LED** off = 0). The test clip, by means of properly spaced teeth (that is, the individual clips), is connected to all pins of the **IC** under test. You read the **LED** for each pin, to see the state of all pins at the same time. (See Figure 8-22 on page 180.)

The simplest way to go, and perhaps the least expensive, is to use a logic probe. This test instrument is a hand-held probe having

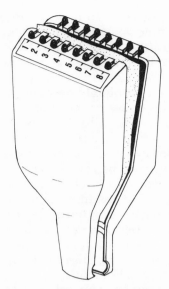

Figure 8-22:   Logic monitor that can be used to check an entire **DIP** 16-pin **IC** at one time. LM-1 monitor from Continental Specialties Corp.

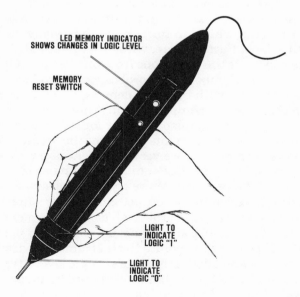

Figure 8-23:   Logic probe. This probe detects and indicates both high and low logic levels in TTL and CMOS digital circuitry. It will also show "bad" logic levels as well as the polarity of a signal pulse as short as 10 nano sec. Example shown is Heathkit IT-7410

one or more indicator lights near its probe tip, as shown in Figure 8-23.An external connection is made to the circuit to be tested (generally V_{CC} and ground).

To use the logic probe, simply touch the probe tip to an **IC** pin or PC card run and observe the indicator light (or lights). There are several different probe manufacturers and light arrangements. Some probes have only one indicator light, thus, if you see an indication (a glowing light), it's a logic 1 and, without light, it's logic 0. Other probes (such as the one shown in Figure 8-23) may have two or more indicator lights. For instance, they may have a light to indicate logic 0 state, another for logic 1, and yet another to show changes in logic level. Also, in some instances, it's necessary to use a separate regulated DC power supply to power the probe, and in others the unit may be powered by the circuit under test.

Generally, you can use a probe to detect pulses of short duration that have low repetition rates. These pulses are sometimes very hard for a general purpose oscilloscope to detect, if not impossible in most cases. However, it should be remembered that all these logic probes are practical only for analyzing certain circuit conditions and, in some cases, only certain **IC's**. If you can afford it, the solution to these problems is provided by two other instruments; the logic comparator and logic analyzer. But, the comparator does have problems in that it may not function properly with certain **IC's** and some circuit connections. In fact, all logic tools have at least one or more conditions where the tester cannot do the job. One example, is open-collector **IC's**. The best way to test circuits such as these (or any open or short circuit), is to use a *current tracer* and logic pulser.

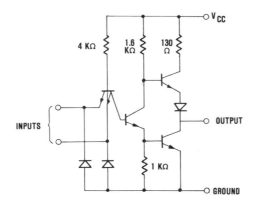

NOTE: COMPONENT VALUES SHOWN ARE NOMINAL.

Figure 8-24:   Input and output circuitry for TTL series NAND gate (Courtesy Signetics Corp., Menlo Park, California)

One type of current tracer looks like a logic probe but works a bit differently. To begin, you don't use the probe tip to test for a voltage pulse in a circuit under test. You use this probe to follow the current flow to a defective circuit. The tip contains a magnetic sensor that is used to monitor the magnetic field produced by the current flow in the circuit under test. Whichever instrument or combination of instruments you use for troubleshooting digital circuits, you should have the technical literature on all of the most common **IC's** (i.e., functional truth tables and pin configurations), especially the ones you service most often. Also, it's a great help in many troubleshooting jobs to know the input and output circuits for the logic families such as **TTL, CMOS,** and **MOS.** An example of an **IC's** input and output circuits for a **NAND** gate of the **TTL** series is shown in Figure 8-24 (this is a single gate from the 7400 shown in Figure 8-9 A).

CHAPTER **9**

Practical Test Equipment Procedures
and Troubleshooting

Getting peak performance from your most often used test gear (for example, your DMM and scope) is what this chapter is all about. Even if you have only a basic knowledge of the solid state circuits used in modern DMM's and what a scope specification should be for digital troubleshooting, you'll find the material in the following pages of immense value—measurable in hours and dollars.

You'll find down-to-earth explanations about the so-called ½ *digits*, how to make low-cost accuracy checks on all types of instruments, analog-to-digital converters (the back bone of DMM's), ohms converters, calibration, and performance checks, with emphasis on troubleshooting, repair, maintenance.

The oscilloscope you use for digital troubleshooting must be capable of presenting complex digital signals in order to determine if the IC under investigation is functioning correctly. However, many oscilloscopes just can't do the job. For example, some of them can't make digital circuit frequency measurements, time measurements, threshold measurements, and the like. In this chapter, you'll find a guide for checking a scope's specifications to see if it will meet your job requirements. If you don't have the right scope, trying to isolate a fault in a digital circuit can be both difficult and time-consuming (if not impossible). Most of us know that a scope with a dual-trace unit and triggered sweep is best, but there are other considerations. You'll find these explained in the last section.

In this chapter, you'll also find an in-depth look at some common test gear faults, troubleshooting procedures, maintenance procedures, and a lot of circuit operation that will add to your troubleshooting speed, increase your knowledge, and, hopefully, improve your personal income.

Some Things You Should Know About
Modern Test Equipment

Almost all electronic test gear circuits are classified as either analog or digital. When you're troubleshooting a particular circuit in a piece of test equipment, it's easy to tell the difference between the two. Analog circuits are the ones where the signal voltages present may be at any level between two extremes (for instance, V_{CC} and ground, usually determined by the instrument's DC power supply). Another "key" is that many of these type circuits are called *linear* because they produce an output directly proportional to the input. Of course, there is a limit, as in all instruments—for instance, an analog multimeter where the meter readings are only correct within the amplitude and frequency range of the meter you're using. Digital circuit operation, on the other hand, recognizes only two signal states, logic high and logic low, as was explained in Chapter 8.

Although it would be nice if we could choose to specialize in either digital or analog circuits, it just can't be done if you're servicing today's electronic instruments. Sure, you'll find an amazing variety of circuits that function using only logic high and logic low, but there are just as many around that use only analog techniques. However, it's equally important to realize that there are many electronic systems that must have both methods incorporated in the design, in order to perform the desired task. One excellent example of combining analog and digital techniques is the digital multimeter. Figure 9-1 is a block diagram of a typical digital multimeter.

If you're reading a schematic for an electronic instrument, one of the easiest ways to determine whether the instrument requires both analog and digital troubleshooting techniques is to look for an analog-to-digital (A/D) or a digital-to-analog (D/A) converter (more about converters later). When reading schematics for instruments such as digital multimeters, you'll find A/D converters located just prior to the display section of the instrument.

Tips for Battery-Operated Test Gear

Many electronic instruments are powered by internal, rechargeable batteries that allow the instrument to operate for at

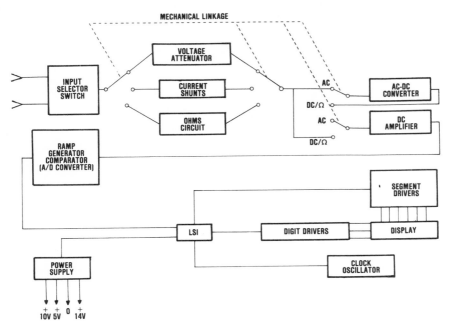

Figure 9-1:   Digital multimeter block diagram. The upper section is analog, the lower is digital: an example of combining both analog and digital techniques.

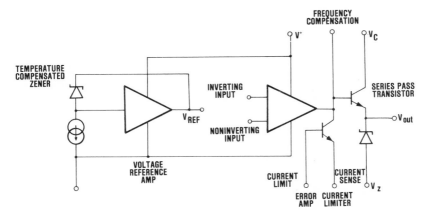

Figure 9-2:   Equivalent circuit for a 723 voltage regulator IC that can be used in a 0.01% voltage reference system

least eight hours. A good rule for recharging is: whenever the light quality of the display is too low to read, the batteries should be recharged. If you're not using the instrument during the recharging period, it usually takes about 12 to 14 hours to bring the batteries back up to full charge. But when recharging and using the instrument at the same time, the recharging time can be extended considerably—up to over two days and nights, in some cases. The point is: be sure the instrument is fully charged before taking it out on a job where no power is available and you will need to use it for an extended period (say, eight hours or more). Also, instruments containing NI-CAD batteries should not be stored for long periods of time without recharging them at least every 90 days. You can help this problem by storing the instrument in temperatures below 25°C (77°F). Another word of caution: when soldering or desoldering, a safe way to go is to remove the batteries or place a piece of insulating material between battery and holder contact. Of course, the power switch on the instrument should be set at off and the AC line cord disconnected.

Practical, Low-Cost Accuracy Checks

All of us need a precision voltage or current reference in our shop. We use these references for calibrating oscilloscopes, voltmeters, ammeters, A/D and D/A converters, and regulated DC power supplies (all of which employ a reference voltage or current).

To even mention *precision references* a very few years ago, was out of the question because they were out of reach of almost all of us, due to cost. Not so today, since with low-cost IC's, good precision references can be built by almost anyone. Take IC voltage regulators, for instance, Signetics µA723 (there are quite a few companies that produce the 723: for example, Fairchild, National, and Silicon) is a linear IC precision voltage regulator that features *0.01% line and load regulation*, with an output voltage that is adjustable from 2 to 37 volts. Figure 9-2 is the equivalent circuit for this monolithic precision voltage regulator IC.

IC voltage references such as these, allow you to build simple voltage and current references that perform as well as all but the best reference previously available. The 723 IC shown in Figure 9-2 contains a temperature-compensated reference amplifier (the amplifier symbol on the left), error amplifier (the amplifier symbol on the right), series pass transistor, and current limiter, with access to remote shutdown. The IC can be wired in two voltage regulator circuits, that is, a low or a high-voltage regulator (low, from 2 to 7 volts, high, from 7 to 37 volts). Figure 9-3 shows a low-voltage circuit and Figure 9-4 shows a high-voltage circuit. Both circuits are courtesy Signetics Corp., Menlo Park, California.

Both of these circuits shown may tend to drift with changes of component junction temperature (particularly when operated under conditions of high power dissipation). Usually, commercial voltage reference standards using IC's like these install the entire circuit in

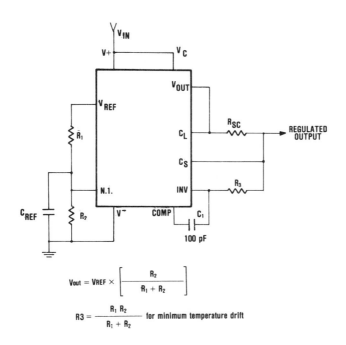

$$V_{out} = V_{REF} \times \left[\frac{R_2}{R_1 + R_2} \right]$$

$$R3 = \frac{R_1 R_2}{R_1 + R_2} \quad \text{for minimum temperature drift}$$

Figure 9-3: Circuit diagram of a low-voltage regulator using a 723 IC

proportional control temperature ovens. However, as shown in the illustrations, by using care when selecting resistors R_1, R_2, and R_3, temperature drift can be held to a minimum. As in all precision voltage reference supplies, it's better if the resistors have a low temperature coefficient. In most cases, this means using wire-wound or metal film precision resistors. Fortunately, actual resistance value is not too critical. So, if your calculation results in some "odd ball" number, generally, you can use the nearest standard value. This will change the R_{REF} value but it probably won't change the output voltage.

There are quite a few **IC's** on the market that can be used as a precision reference; for example, National's LM 199. This **IC** temperature stabilized zener diode uses an on-chip heater. You can use this **IC** just as you would any zener diode so far as circuit design is concerned. The difference is in it's excellent thermal stability. You do have to have a 9 to 40 volt DC power supply to connect to the heater pins. This four-pin zener diode and its heater, are shown in Figure 9-5. In general, it's best to ground pins 2 and 4 so that the internal substrate will remain reverse biased.

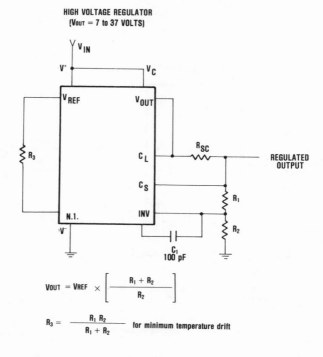

Figure 9-4: Circuit diagram for a high-voltage regulator using a 723 IC

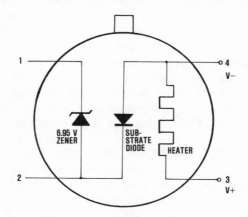

Figure 9-5: A temperature-stabilized zener diode (National LM 199) that can be used for a precision reference voltage

From time-to-time, a reference current rather than a reference voltage, is needed in almost any shop. You'll find that the 723 **IC** can also be used as a constant current source, and that you can purchase constant-current diodes, plus several other devices that can be used as a reference. These are available from your local electronics supply store. You might try Motorola's MC 1404X and MC 1504X or, possibly, an REF 01 or 02 voltage regulator could be used for the reference job you have in mind.

Troubleshooting Digital Multimeters

Like everything else in the electronics field, the humble multimeter has undergone a shrinking process. We had to troubleshoot transistors (in large instrument cases) in the 50's and 60's, then the **IC's** began to show up in the 60's and 70's, and the trend toward smaller and smaller instrument packages began. The biggest difference was made possible by the meter-on-a-chip **IC's** such as the **CMOS** analog-to-digital converter, ICL 7106 and 7107. Today, you'll find all the digital circuitry for multiple-decade counters, display and logic circuits (not including the actual displays), and oscillators, reference, and analog-to-digital converter components (there are a few external components), *all on a single chip.*

Because it is necessary to troubleshoot pre-1976, 1977 instruments, let's first look at a digital multimeter of about 1972 vintage. A block diagram of a Fluke model 8000A DMM is shown in Figure 9-6. In this instrument, the input circuits are a collection of resistors, switches, a few capacitors, and diodes. Troubleshooting these circuits is very much the same as what you have always done with standard shop multimeters.

The blocks following the input stages (see Figure 9-6) are the AC converter and ohms converter. Basically, AC converters consist of a transistor (Q_1), operational amplifier (U_1) and a rectified voltage smoothing filter (see Figure 9-7). When troubleshooting, remember that the transistor (Q_1) acts as a buffer for the **OP AMP**. Next, notice that the output of the buffer is fed into the non-inverting input of the **OP AMP**. Because of negative feedback, the inverting input will follow the non-inverting input. In turn, a current will flow through the diodes and cascaded components to ground. The diodes conduct on alternate half-cycles, resulting in a rectified voltage, which is filtered and available at the output. The output voltage is the DC voltage required for the A/D converter.

The ohms converter in instruments of this type, is even

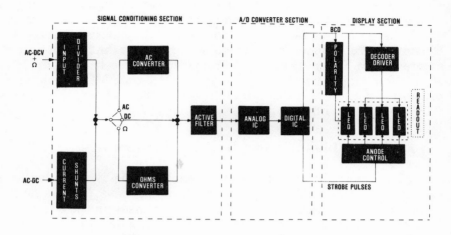

Figure 9-6:   Block diagram of a Fluke Mfg. Co., 8000A digital multimeter.
See text for troubleshooting information pertaining to
instruments such as this

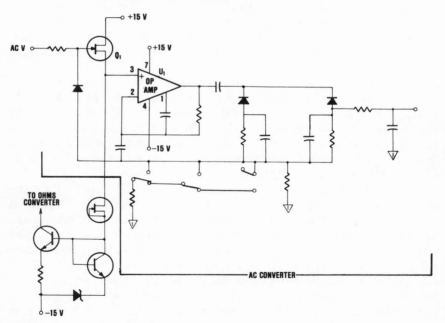

Figure 9-7:   AC converter schematic for a digital multimeter (model
8000A). The AC input voltage is rectified, filtered, and
produced as a DC output used to drive an analog-to-digital
converter

simpler. The entire operation uses an **OP AMP**, a couple of transistors, a couple of diodes, and a few resistors and capacitors, as shown in Figure 9-8. The reason for the AC and ohms converter is that the **IC's** in this, and similar, instruments must have a DC input to the analog **IC**. Therefore, troubleshooting the circuits up to the **IC's** in a DMM such as this requires no more than the standard gear you already have in your shop, plus the servicing information presented in Chapter 1. You'll find the essentials of **IC** troubleshooting in Chapter 2, and Chapter 8 covers the problems of checking digital **IC's**.

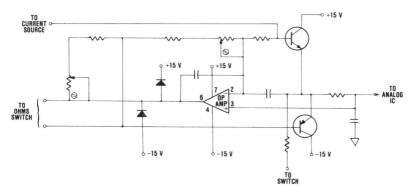

Figure 9-8:   Ohms converter circuit for a DMM

There are a few points that should be emphasized. The **IC's** in the DMM we are discussing (and all others that I know about) use external components. The first step in troubleshooting should be to check that these components are working correctly. Because these components are usually capacitors, resistors, zener diodes, etc., another multimeter will probably be all you'll need to check them. Of course, it's understood that you should have the instrument's service notes.

You'll find that in an **OP AMP**, one single resistor in the feedback line changing value can change the entire characteristic of the amplifier. So, it's very important to check these components. Next, as in any circuit, check the power supply voltage at the proper terminal of each **IC**. A fast and easy way to determine if the **IC** is drawing too much current is simply to touch the **IC**. If you can't keep your finger on it for at least a few seconds, it is too hot! If you've ever troubleshot tube or transistor circuits, you know that this is an old trick and, as with these circuits, "too hot" means you either have a bad **IC**, or some associated part is defective. By the way, digital **IC's** are usually operated from one or two fixed voltages (the 8000A uses two, +15V and −15V) and, if these voltages are within 10 percent of your meter readings, you can usually asume that the **IC** is not troubled by the supply voltage. However, this is not true of the analog **IC's** found in these, or any other, instruments. Be sure the pin voltages are right on the money when troubleshooting analog **IC's**, particularly if the **IC** is using a reference voltage, bias voltage, or a control voltage.

Today's trend, as we have said, is toward meter-on-a-chip. Several companies such as Intersil, Motorola, National Semiconductor, RCA, and Signetics, are producing these chips. Incidentally, in some cases, these are **IC** sets, rather than a single chip. One such instrument that uses a meter-on-a-chip **IC** is Hickok's model LX 303. They used an ICL7106 to build their DMM. The instrument is about the size of a small hand-held pocket calculator (weighs only 12 ozs) and looks like the one shown in Figure 9-9.

To see how easy it is to troubleshoot a modern DMM like the one shown in Figure 9-9, let's take a look at a simplified version of the instrument's schematic (see Figure 9-10). First of all, as you can see, the DMM circuitry is based on the ICL7106. The display unit is a 3½ digit liquid crystal display.

Let's digress for a moment and talk about ½ digits. In general, each digit in a read-out (**LCD** or **LED**, etc.) can be any number from 0 to 9. Therefore, if we have a 2-digit DMM, it can give us a reading of any number from 0 to 99, or decimal fraction such as 2.2, 9.9, etc., depending on the number of ranges the instrument has. But, by adding the "half" digit—in reality, a 1—the instrument can be made to display up to three significant figures (there are exceptions, but in most cases you'll read three). In other words, a two-digit instrument can read 9.9 but a 2½-digits can read 1.99, and so on (for example, 2½-digits, 1.99, 3½-digits, 1.999). The manufacturers have found that the cost of adding a ½-digit circuit is less than adding a full 0 to 9-digit circuit. Hence, the reason they like the ½-digit circuits.

If you like to experiment, National Semiconductor Corp., 2900 Semiconductor Drive, Santa Clara, CA. 95051, *did* (at least they did in 1978) have a very good data bulletin on using their single-chip **IC** ADD 2500. You'll also need a 2½-digit **LED** read-out. They suggest either a NSB 3881 or NSN 333, or anything similar, to complete the meter circuit. Other than these two units, you'll only need a few capacitors, resistors and, of course, a power supply, to construct your own instrument.

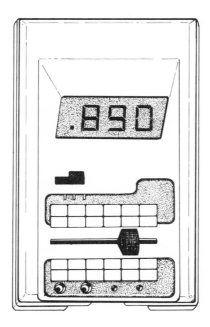

LOW VOLTAGE REGULATOR
(Vout = 2 to 7 VOLTS)

Figure 9-9:   Example of a modern DMM. Hickok Co., Model LX303. Dimensions and weight: 5⅞" × 3⅜" × 1¾", 12 oz, power, 9V battery

Now, back to troubleshooting Hickok's DMM. First, you should refer to the schematic shown in Figure 9-10. Notice that almost every component in the circuit can be quickly checked with a voltmeter. The precision resistor network in the meter input circuit is really a network of laser-trimmed resistors on a single substrate.

It should be noted that, in some cases, certain components are not included in the operating circuit unless the function switch is set to a particular mode of operation. For example, R 107, 108, 113, and the three diodes CR 102, 103, and 104 are only in the circuit when the meter is set in the ohms measuring mode. Therefore, the switch position on this meter, as well as any other, should always be examined to see if it is in the correct position for the particular circuit you're troubleshooting. The LX303 (see Figure 9-9) has a total of 19 ranges and functions.

To isolate a faulty circuit, switch the meter in to each mode of operation and watch the read-out. For example, if you can't make a

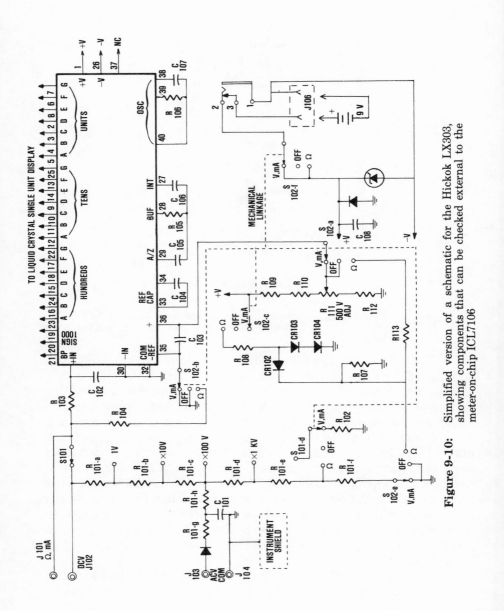

Figure 9-10: Simplified version of a schematic for the Hickok LX303, showing components that can be checked external to the meter-on-chip ICL7106

measurement in an ohms position, check the banana jack plugs, switch S101, and try other ohms ranges. Referring to the schematic of this meter, you'll see several ohms (Ω) terminals on switch 102 (a, b, c, etc.). Each of these switch terminals can be used as a key to which circuit may have a faulty component. Incidentally, this entire meter circuit is mounted on a single PC board. However, although it is small, it isn't cluttered, which means you shouldn't have too much difficulty troubleshooting the instrument. As always, check the **IC** pins for shorts and correct voltages. For high-speed troubleshooting, you should have the DMM's service data. If you don't have this information, try to track it down. For an **IC** data sheet on the ICL 7106 or 7107, or similar meter-on-a-chip **IC**, contact one of the manufacturers listed at the beginning of this section.

To repeat, to isolate the problem to as few circuits as possible, use the controls and displays provided on the front panel of the instrument you're troubleshooting. Try to get all service information, technical bulletins, and data sheets possible, and, above all, *read, look, and think before* starting to troubleshoot any electronic equipment, *especially test instruments*.

General Maintenance and Calibration

It's true that electronic test instruments are, in general, growing smaller and less complex. The new DMM's are very easy to calibrate. There are only three adjustments on Fluke's model 8020A and just one on Hickok's model LX303. But, the manufacturers have not lost sight of the fact that most test gear is used daily on the bench and must stand quite a bit of wear. What this means is that you won't have much trouble with calibrating or breaking the instrument cases used on today's test gear. These cases are usually plastic, but rugged; for instance, the case on the LX303 is guaranteed for one year. A few words of caution: 1) because of the plastic cases used today, it's very common to find shielding used inside the instrument case (the purpose is to prevent stray electromagnetic fields from confusing the high-impedance **CMOS** inputs of the **IC's**); 2) all ground connections to instrument shields are important for proper operation; and 3) clean a plastic case with denatured alcohol or a mild solution of detergent and water. Do not use aromatic hydrocarbons (a solvent having a distinctive smell and using compounds such as benzine and methane) or chlorinated solvents (example; household bleach) because they may damage the plastic materials of the instruments.

The test equipment you'll need for performance tests and calibration of most modern DMM's and the like, includes a DC voltage source, DC current source (see "Practical, Low-Cost Accuracy Checks," in this chapter), AC voltage and current source, a few resistors ±0.1%, and a frequency counter that will measure 100 m sec pulses with at least 1 μ sec resolution. You'll find additional information for digital equipment servicing in Chapter 8.

When you make performance checks on an instrument, it is important to let the instrument warm up to room temperature. It has been suggested that room temperature be set at 68°F, to save energy. However, although this is an excellent suggestion if you're trying to save energy, it's not a good temperature to use during performance checks and calibration. For best results, the environmental conditions surrounding the instrument should be maintained between 72°F to 77°F (22°C to 25°C), with a relative humidity of 70%. You'll find that manufacturers may give other temperature and humidity ranges, but if you don't know what they should be, the ones given here may be used, in most cases. The symptom of a too cold or too hot environment will generally show up as a condition where it is impossible to adjust to the manufacturer's specifications.

If you're working with a digital read-out instrument, it must be remembered that a "zero check" will not be the same as with a standard analog meter read-out. You can set an analog meter reading to exact zero on the meter face (using both mechanical and electrical adjustments). However, this is usually not true with a digital read-out. For example, to make a "zero check" on the 8000A digital multimeter we mentioned before, with the instrument energized, you depress the push buttons labeled DCV and 200 MV. Next, short the ohms terminal to the common terminal, and the read-out should show some number quite a bit less than a whole digit. Remove the short (in other words, open circuit), and the read-out should indicate some value quite a bit less than 10 digits.

Generally, you'll find that display limits are given for the various function/range settings on the instrument. Again, using the 8000A as an example, an AC current performance check with a 200 μA range setting, and an input of 190 μA at 100 Hz, should produce a read-out within the limits of 186.1 to 193.9. Notice, if you have the service manual for a certain digital read-out instrument that lists the display limits, it's fairly easy to check each function/range setting for a trouble. Simply apply the required inputs, check and/or adjust for in-limits indications and the one that won't come in is the circuit you should start troubleshooting. Calibration is just as simple. Make a "zero check" (something like 1 to 3 internal adjustments on modern instruments), using a scope, frequency counter, etc., and set the counter to within the specified time variations and, most of the time, that's all there is to it.

If you're trying to calibrate an instrument using a meter read-out, in most cases an approximate calibration can be made at the center and both ends of the scale by following the service manual instructions and adjusting the internal screw driver adjustment. If your requirements aren't too exacting (you can usually get away with at least 5% error during most troubleshooting measurements), a new battery and 1% wire-wound resistor can be used to calibrate ohmmeter scales of low-cost analog shop instruments. Oscilloscopes usually can be calibrated using 60 Hz power line frequencies (tolerance of power line frequency is maintained within ±0.02 Hz), and observing the waveforms produced on the scope. It is *strongly* suggested that you use a step-down transformer—say an old 6.3 filament transformer—before you connect the power line to the scope.

You can use the same setup just explained, to check a signal generator. Simply connect the output of the signal generator to the scope vertical input and the output of the step-down transformer to the scope horizontal input. When you have one single sine wave on the scope, both inputs are exactly the same frequency, as we all know. Two cycles are a 2 to 1 ratio, three cycles are a 3 to 1 ratio, and so on. As you'll remember, it depends on which inputs you connect to the scope and which frequency source is the highest or lowest. Of course, both scopes and signal generators can be calibrated by using any of the counters available today.

Checking the Analog-to-Digital Converter

Generally, any digital equipment servicing data will provide schematic drawings that show test points and voltage values and, sometimes, key wave forms will also be shown. With this data, all you need is a DMM, scope, and/or digital probe or clip, and some form of signal generator for a test signal source (depending on whether you're checking analog or digital input circuits). *Note:* when identical **IC's** are used in an instrument, or you have an identical instrument in the shop (assuming the instrument is out of its case), it is far easier, quicker, and generally a more positive check, to interchange the **IC's** and make a cross-check than it is to hook up scopes, clips, and probes, etc.

When you're troubleshooting an A/D converter, you'll find that many instruments use a voltage-to-frequency conversion technique in this unit. You should measure a DC voltage at the input of the **IC** and a certain frequency on the output. This frequency is characteristic of the magnitude and polarity of the DC input voltage. Counting of the output frequency of the A/D converter **IC** is usually done by the digital **IC**.

To give you an idea of what you should expect on the output of an A/D converter when troubleshooting, let's say you are checking Fluke's 8000A digital multimeter. In this case, using a frequency counter, you should measure very close to 80 kHz (the rest frequency) with the DMM energized and no DC signal voltage into the A/D converter **IC**. Insert a positive or negative voltage of a certain amplitude and the **IC** will generate an output frequency above or below the rest frequency in respect to the magnitude and polarity of the input voltage. The output should be capable of varying ±40 kHz from the 80 kHz rest frequency. Actually, the output from the analog **IC** alternates between the rest frequency during one period of time (at a 120 millisecond rate), and a frequency corresponding to the A/D converter input voltage during the next period, and the digital **IC** counts the difference between the two frequencies to provide a binary coded decimal signal.

While there are a number of ways by which A/D conversion can be accomplished, we will limit our discussion to the most popular. Also, because probably the most common use of the A/D converter is in digital multimeters, where, as has been explained, an analog voltage must be converted to a digital signal to drive the circuits that you ultimately see as a numeric display, we'll stick to this instrument for our examples.

One design that is very popular in digital multimeters is dual-slope conversion. The reason for his scheme's popularity is its simplicity. The dual-slope converter is an integrating A/D converter in which the unknown signal is converted to a proportional time interval, which is then measured digitally. In this system, an integrator is used as an input to a comparator. Therefore, let's first examine the basic operation of these two circuits before going into how they are both used in more complex circuits.

In both instances, these circuits are usually designed using **OP AMP's**. You'll find considerable information pertaining to servicing **OP AMP's** throughout this book, particularly in Chapter 5. However, at this point we are interested in the type **OP AMP** circuits that you'll find when troubleshooting digital multimeters, so let's examine some basic circuits. First, Figure 9-11 shows a simple **OP AMP** integrator that is basically nothing more than an **OP AMP** with a capacitor in its feedback circuit, and some form of switching arrangement.

In reality, the switches S1 and S2 are usually diodes, bipolar transistors or field-effect transistors. The switching is normally controlled by a clock circuit. To see how the simple circuit in Figure 9-11 works, let's start off by assuming switch S1 is closed. What happens during this period of time—reset (1)—is that the capacitor (C) in the **OP AMP's** feedback path will discharge, bringing the

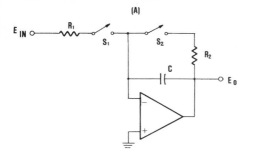

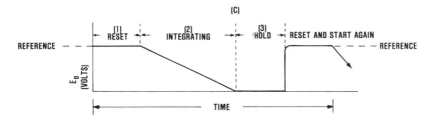

Figure 9-11: Basic OP AMP integrator. (A) is a simplified switching circuit, (B) shows switch positions for each time period, (C) is a graph of the voltage output during each time period labeled reset, integrating, hold, and finally, back to reset

output voltage (E_o) to zero, as shown in Figure 9-11 (C). Next, say that S1 is opened and S2 is closed. During this period of time—integrating (2)—the capacitor (C) in the **OP AMP** feedback loop will rapidly change to some voltage (say a negative 4 or 5 volts), which will appear on the output (E_o). Just as the output voltage during the integrating period (2) reaches its lowest value, switch S2 opens. This places the voltage buildup on the capacitor—shown as hold (3)—at this fixed value until S1 is again closed, returning the output voltage back to zero and starting a new cycle on the next trigger pulse.

Now that we have a general idea about integrator operation, here are a few words about the operation of an **OP AMP** being used as a comparator. To put it simply: a comparator circuit is one that provides an indication of the relative state of two input potentials (see "Differential Amplifier," Chapter 5). When these circuits are used in instruments such as DMM's, you'll usualy find that one input is a reference potential and the other is an unknown, i.e., the voltage input to the test gear.

As an example of how an integrator, comparator, clock (precision pulse generator), AND gate (see Chapter 8), electronic switching circuit (diodes, **FET's**, bipolar transistors, etc.), and a counting display section (see Chapter 8) might be used in a dual-slope converter in a DMM, see Figure 9-12. This setup is, perhaps, the

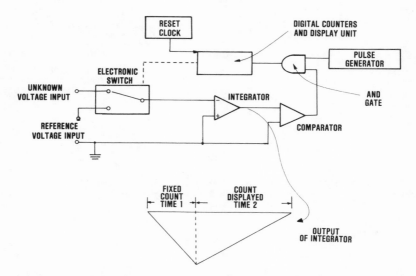

Figure 9-12: Block diagram of a dual-slope converter frequency used in many types of digital multimeters. The graph of the integrator output assumes V_{in} is an analog voltage at a fixed level and the counter circuits have just been reset to zero (starting point)

easiest to understand. It is shown much simplified and is basically a ramp voltage comparison system.

In many cases, you'll find that the counters used in these systems have 1,000 possible states. Now, as you can see, the switching circuit (S1) connects the analog voltage being measured to the input of the integrator, which starts a capacitor charging (see Figure 9-11). The linear charge voltage on the capacitor continues to change until the counter circuit has completed its 1,000 counts. Just at the end of the fixed count time, the switching circuit should switch the input to the integrator over to the reference voltage. This is time T_1, shown as the output of the integrator.

Next, with the reference voltage on the integrator input, of course it has to integrate this voltage (which it does) until it reaches zero volts, as shown on the second slope on the graph (T_2). What happens next is that at time T_2, the comparator switches states and turns off the clock pulse generator (this is done by the AND gate), and the counter remains at whatever count it contains, waiting for a new reset pulse to start the entire cycle over again. Notice, it takes both time T_1 and time T_2 for a complete conversion of the analog signal to a digital count. Hence, the reason this system is called a *dual-slope converter.*

The A/D converter is the backbone of most digital read-out measuring equipment, therefore, when troubleshooting, it's a good idea to know what's inside one of these chips. Figure 9-13 shows the internal, and some of the external, connections of Siliconix's LD111 and LD110 **IC**. The Siliconix Co. is located at 2201 Laurelwood Rd., Santa Clara, CA 95054.

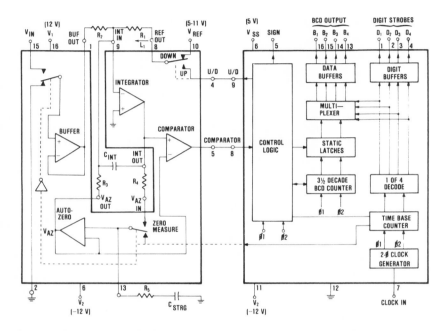

Figure 9-13: A/D converter using two of the Siliconix Co. IC's (LD110 and LD111)

When you're working with A/D converters, many times you'll find that signal conditioning circuits have been included. Three of the most common are; *preamplification* (the preamp will frequently be another **OP AMP**), *filtering* (to remove noise components from the input signal), and some form of *isolation* (to eliminate noise pickup, ground loops, etc.)

A Guide to Checking an Oscilloscope's Specifications for Modern Equipment Servicing

Because today's troubleshooting demands much more from an oscilloscope, it is extremely important that you understand the numerous specifications dealing with the scope's capabilities. For example, *vertical deflection sensitivity* and *vertical bandwidth* are two of the most important specifications given for an oscilloscope. In fact, both of these constrictions can prevent you from making certain measurements.

Vertical Deflection Sensitivity: The deflection sensitivity of a certain scope is telling you what the smallest voltage is that will produce a 1-centimeter deflection on the CRT. *Note:* most all modern scopes use a per-division (generally, 1 cm) specification, however, some older scopes use the inch, or full-screen deflection, as a standard. The vertical deflection sensitivity of Heathkit's model 10-4235 is 2 mV/cm to 10V/cm. In comparison, their model 10/SD-4235 input sensitivity (same thing as deflection sensitivity) is 10 mV/cm to 20V/cm. Notice, the first scope will measure right on down to 2 mV/cm and the second one, only down to 10mV/cm. So what's the difference? Well, if you're working with very small signal amplitudes, your best bet would be the first scope, but there is, as always, a catch. The first scope costs just about twice as much as the second one. On the other hand, if you're operating a small home shop and have a limited budget, or like to experiment, the second scope may well be all you need.

Vertical Bandwidth: Basically, the same applies to the vertical bandwidth specification of a scope. For example, Heathkit's model 10-4105 (a comparatively inexpensive scope) vertical bandwidth specification is DC to 5 MHz. What does this really mean so far as what you can do with the instrument? To answer this, you should understand how vertical amplifier bandwidth is defined. The vertical bandwidth of an oscilloscope is defined as the point at which the signal on the CRT has been reduced by 3 dB, with respect to the low frequency reference point (half-power, or 0.707 signal voltage, in respect to the voltage reference).

As the vertical signals increase in frequency, there should be a continuous decrease (roll-off) at a rate slightly greater than 6 dB per octave (the interval between any two frequencies having a ratio of 2:1). If the roll-off is at a much greater rate than this, you're not going to have very accurate reproduction of a high-frequency complex signal. Furthermore, a signal that is reduced in amplitude by 3 dB due to an increase in frequency, will suffer because of a large phase shift in respect to the reference point (about 45°), which may cause a serious error in some of your measurements.

As we said before, one of the deciding factors is cost. You can purchase an oscilloscope with a vertical response using DC coupling (DC coupling will usually produce slightly better bandwidth than AC coupling), of DC to 35 MHz. But it will cost about three times as much as a low-cost DC to 5 MHz scope.

Vertical Rise Time: Another specification you'll find important is vertical rise time. A couple of examples of vertical rise time are 70 nano seconds (not the best) and 10 nano seconds (very good, for most uses). Rise time is defined as the time required for the signal you're measuring to increase in amplitude from 10% of its total value to 90% of its total value on the CRT of the scope (see Figure 9-14).

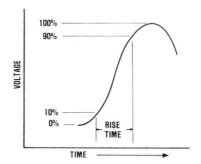

Figure 9-14:   Both rise time and fall time of a scope are measured using this definition.

Rise time is especially important to you if you're going to use the scope for pulse analysis. If you're going to do a lot of this type work, a good rule of thumb to use when you choose a scope is: the scope should have a rise time that is equal to, or less than 20% of the rise time of the fastest rise time pulse you expect to measure.Notice (using this rule), you still can't see fast digital pulses using the 10-nano-second rise time scope that was mentioned as being very good. However, a logic clip or probe will generally take care of scope problems such as these.

Vertical Input Attenuator: For some measurements that require full-screen displays, the scope input attenuator becomes important. Try to adjust an "oddball" amplitude signal to exact full-scale on the CRT and some scopes just can't do the job. The best way to get around this problem is to be sure the scope has a variable control that adjusts the effective attenuation between the indicated value and its next highest position. Let's take Heathkit's model 10/SD4510 as an example of such an instrument with an adjustable vertical deflection system. The specifications are; sensitivity 2 mV/cm to 10V/cm, 12 steps in 1-2-5 sequence, variable continuous between steps to approximately 30V/cm. Accuracy of attenuation is normally ±3 to ±5%, depending on temperature, in reference to a certain deflection (such as 1V/cm).

Vertical Delay Line: Although a scope with a vertical delay line usually costs more, it is essential that a delay line be included if the scope is going to be used to analyze digital circuitry. The specification that most of us are interested in will indicate the number of nano seconds of the pre-triggered waveform that you'll see on the CRT. For example, Heathkit's model 10-4235 scope delay line allows display of at least 20 nano seconds of pre-triggered waveform. This pre-trigger observation on the CRT is especially necessary when you're working with digital circuits, where the measurement of pulse rise time is frequently necessary to determine if the pulse is correct for proper operation of a particular circuit.

Trigger Sensitivity and Bandwidth

It would be nice if all manufacturers used the same method of defining trigger sensitivity and triggering bandwidth but, unfortunately, they don't. However, in general, you'll find that trigger sensitivity will indicate the smallest deflection that you can produce with an input signal and still hold a stable wave pattern on the CRT. A good rule is: a vertical trigger sensitivity of one division (in almost every case, 1 cm), or better, can be considered as good. If you can't get a wave pattern to hold stable unless the waveform occupies more than 1 cm on the CRT, it's suggested that you check your job requirements to see if the scope has sufficient trigger sensitivity for your needs.

Trigger bandwidth may also be important to you. Perhaps the best definition of this parameter is: the highest frequency at which you can maintain a stable pattern on the scope CRT, with the manufacturer's specified value (most of the time, 1 cm). The reason trigger bandwidth may be important is that it tells you how the scope will perform (trigger) on complex waveforms and how stable you will find the scope to be when working with high-frequency signals. For very good triggering, you would want a scope that has a triggering bandwidth about twice the scope's vertical bandwidth. However, in many service jobs, you can get by with less than this, although a triggering bandwidth of less than the scope's vertical bandpass will probably cause you quite a bit of trouble when, and if, you try to observe various complex waveforms.

Volts-Per-Inch Vs Volts-Per-Centimeter

Some scopes are rated using volts-per-inch, and many are rated using volts-per-centimeter, when referring to vertical amplifier sensitivity. What's the difference? Quite a bit, if you're paying out your money. One scope may look better, but when all the facts are in, they may end up exactly the same. For example, not one of the following scopes (see Table 9-1) is better than the other, it's simply the way the manufacturer presents the information: rms volts, peak-to-peak volts, and whether it is based on inches or centimeters. In other words, it's best to check out a scope's ratings very carefully. You could spend more money for less performance, simply because the specifications *look better*. See Table 9-1 for a comparison of four scopes with exactly the same rating except for the method of presentation.

1	RMS VOLTS PER INCH	0.01
2	PEAK-TO-PEAK (OR DC) VOLTS PER INCH	0.028
3	RMS VOLTS PER CENTIMETER	0.004
4	PEAK-TO-PEAK (OR DC) VOLTS PER CENTIMETER	0.011

Table 9-1: Four oscilloscopes having exactly the same sensitivity rating except for the method of presentation. Remember, an inch contains 2.54 centimeters. Therefore, scope number 1, having a sensitivity of 0.01 rms volts-per-inch must have an equivalent sensitivity to scope number 3 of 0.004 rms volts-per-centimeter

CHAPTER 10

A Brief Review of Some Important
Digital Fundamentals

This chapter will give you a functional understanding, along with troubleshooting procedures, for several important digital circuits. You can use the information in the following pages to help diagnose problems with logic circuits, analyze flip-flops, and understand digital pulses (for instance, one section explains the three basic flip-flops with logic tables that can be used during troubleshooting).

A practical explanation of binary words, bits, byte, nibble, and binary codes is also included. You'll find that working with digital technical data is more rewarding, if you understand Boolean Algebra. This subject is made easy in the third section of this chapter.

The basic building block of digital IC's is the gate. When gates are multiplied and combined, they provide most of the essential characteristics of digital IC circuits. You'll find many examples of practical combinations of the basic AND and OR gates in this chapter.

Numbering Systems

The decimal number system (using base 10 to perform arithmetic operations) is used in the world outside of digital equipment. Inside digital circuitry, the *binary number* system is most often used. For example, using the common decimal system, we can calculate $5 + 5$ and know the sum is 10. But, the binary system uses combinations of 0's and 1's to represent and perform the same operation $(101 + 101 = 0101)$. In other words, the binary number system uses base 2. An example table using base 2, is shown in Table 10-1.

2^7	2^6	2^5	2^4	2^3	2^2	2^1	2^0
128	64	32	16	8	4	2	1

Table 10-1:   Conversion chart that can be used to change a binary number to a decimal system number

You can extend the table to the left indefinitely by doubling the last number in the bottom row. Notice, the last number on the left is 128. The next should be $256(2^8)$. To change any binary number to a base 10 number, simply draw up a table with the last number on the left as high as needed, then write the binary number underneath the table and *add* the decimal equivalent where the binary 1's appear (the 0's aren't important in this procedure). For example, Table 10-2 shows the binary number 01010100 placed under the conversion table. Now remember, add up only those numbers that appear above a binary 1. In this cae, $64 + 16 + 4 = 84$.

2^7	2^6	2^5	2^4	2^3	2^2	2^1	2^0
128	64	32	16	8	4	2	1
0	1	0	1	0	1	0	0

Table 10-2:   An example of using a conversion chart to change the binary number 01010100 to its equivalent decimal number 84 (read the chart from right to left)

Binary Words: Referring to Table 10-2, you'll notice that there are eight binary digits (0's and 1's). Such a combination of eight binary digits is commonly used in equipment such as microprocessors to form a binary *word*. Binary digits are referred to as bits (formed from the two words *binary* and *digits*, using *bi* and *its*). Therefore, an 8-bit system is one in which eight binary digits are used to form a word. In an 8-bit system, it is always necessary to use the full eight bits to write a binary number.

When all eight bits are used at once, it's often referred to as a *byte*. However, there are many times where less than eight bits are required to express a number. In an 8-bit system, four bits are generally called a *nibble*. But you'll find that other amounts of bits are referred to as a nibble. For instance, there are many computers that will accept either 16-bit or 8-bit binary words. Anything less than a complete word may be called a nibble.

Binary Codes: Although there are quite a few different codes based on the binary system, the one that is the most widely used is known as *binary-coded decimal* or, more commonly, **BCD**. In this system, decimal 1 is represented by 0001, decimal 2 by 0010, and so on. However, if a decimal number has more than one digit, you'll find that four binary bits are used for each decimal digit. For example, the decimal number 3231 is represented by 16 binary digits in groups of four. See Table 10-3.

DECIMAL	3	2	3	1
BCD	0011	0010	0011	0001

Table 10-3:   Binary coded decimal. The decimal numbe 3231 represented by 16 binary digits

If you are troubleshooting a typical 8-bit system such as a microcomputer, you'll find each byte can be viewed as containing two 4-bit **BCD** numbers. In digital equipment, you'll find that all circuits operate using binary. In most equipment such as microcomputers, you'll find that the codes used external to the electronic circuits use other codes; for instance, hexadecimal and ASCII. Conversion between the various codes is done by decoders, encoders, and code converters.

One thing to keep in mind when servicing digital equipment is that all systems do not use the *American Standard Code for Information Interchanges* (or ASCII). In other words, you may encounter other types of keyboards or printouts. But whatever the code, you'll find that binary is used inside the equipment and, in almost every case, it will be **BCD**. This is because binary is most compatible with the signal pulses used in digital circuitry, as has been explained in previous chapters (see Chapter 8).

To troubleshoot a single chip such as those shown in Chapter 8, is fairly straightforward, if you have a truth table and other service data. But, if you are going to troubleshoot microprocessor systems, you must have operational instructions before the computer will perform any function. This information is provided by the manufacturer and is generally known as the *Instruction Set.*

Analyzing Logic Circuits

The mathematical tool for analyzing the operation of logic circuits is known as *Boolean Algebra.* A Boolean equation that you might find in a manufacturer's data book or application note may look like this:

$$3 = (4+5) + (6+7)$$

where each number represents a pin on a particular **IC** gate. In the strictest sense, Boolean equations are in algebraic form, i.e., A, B, C, etc. However, in the real world, we generally work with **IC** pin numbers. Therefore, we'll use pin numbers rather than alphabet letters.

Now, let's see what the equation is telling us. First, the expression indicates that there are two OR gates being OR'd by another single OR gate. Next, let's draw the symbol for the first part of the equation (pin 4 + pin 5). This is shown in Figure 10-1.

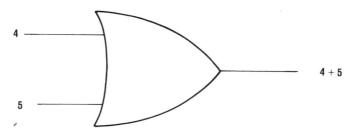

Figure 10-1: First part of the Boolean equation (see text) for an OR gate (using IC pin numbers)

Next, draw the symbol for the second part of the equation (pin 6 + pin 7). This is illustrated in Figure 10-2, which, as you can see, is exactly the same drawing except the inputs to this OR gate are on different pins of the **IC**.

Finally, we need one more OR to combine the two OR gate outputs (Figures 10-1 and 10-2). How do we know this? Because the + between the first and last part of the equation is required to produce

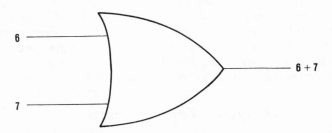

Figure 10-2:   Second part of a Boolean equation (see text) for an OR gate (using IC pin numbers)

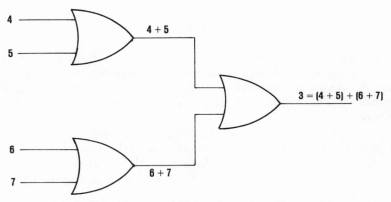

Figure 10-3:   Completed logic diagram of an IC containing three OR gates, using Boolean algebra to describe its operation

the output on pin 3. Combining all three OR gates results in the logic diagram shown in Figure 10-3.

Next, let's say that you encounter the logic diagram shown in Figure 10-4. Notice that its diagram has two OR gates and one AND gate. The AND gate is doing the combining of the OR gate outputs, in this example. The Boolean algebraic expression for the **IC** circuit configuration follows the illustration.

First, we need the equation for an OR gate, which is (3+4). Of course, the other OR gate is (5+6). Now, because the AND gate is doing the combining, the total equation becomes $2 = (3+4) \bullet (5+6)$.

Finally, a more complicated **IC** is Signetic's 10121. This **IC** is a 4-wide, 3-3-3-3 input. OR-AND/OR-AND INVERT gate. Pin 10 is common to two of the gate inputs. This function is particularly useful in data control and multiplexing. Figure 10-5 shows the logic diagram for the 10121.

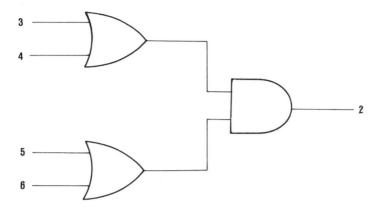

Figure 10-4: Logic diagram of an IC containing two OR gates and one AND gate

The equations (positive logic) for the logic diagram are:

$$2 = (4+5+6) \bullet (7+9+10) \bullet (10+11+12) \bullet (13+14+15)$$
$$3 = \overline{(4+5+6)} + \overline{(7+9+10)} + \overline{(10+11+12)} + \overline{(13+14+15)}$$
$$= \overline{(4+5+6) \bullet (7+9+10) \bullet (10+11+12) \bullet (13+14+15)}$$

In this equation, we see that there are three inputs to each OR gate and the outputs of the OR gates are fed to an AND gate, with reference to pin 2. You'll also notice that the logic diagram shows the inverting symbol (the small circle on the OR gate outputs). The solid bar over $3 = \overline{(4+5+6)}$ etc., denotes inversion.

Flip-Flop Circuit Basics

Flip-flops are primarily used for memorizing (i.e., holding) logic levels and for counting. Over the years, this circuit has had several names; *bistable multivibrator, Eccles-Jordan circuit*, and *trigger circuit*. The flip-flop is often referred to as a trigger circuit or toggle because it has two stable states (called *set* and *reset*). It will remain in either set or reset until its state is changed by an external change in logic level.

There are three basic types of flip-flops that you may encounter during servicing. Troubleshooting logic symbols are shown in Figure 10-6. Incidentally, you'll find that the flip-flop can be easily checked by using either a voltmeter or scope. Simply check

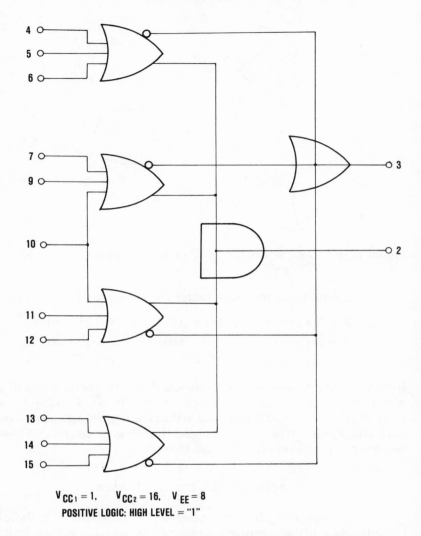

$$V_{CC_1} = 1, \quad V_{CC_2} = 16, \quad V_{EE} = 8$$

POSITIVE LOGIC: HIGH LEVEL = "1"

Figure 10-5:   Logic diagram for a 16-pin DIP Signetic 10121 OR-AND/OR-AND INVERT gate

the state of the output and then refer to the truth table (or manufacturer's application notes, etc.).

The logic symbol for an RS flip-flop is shown in Figure 10-6 (A). To check one of these devices, you would place a logic pulser on either S or R, because they are the inputs. Your voltmeter, scope, etc., would be placed on Q or $\overline{Q}$. These are the outputs. If you place the signal input on S and inject a logic level 1, you should see a logic level 1 on Q and a logic level 0 on $\overline{Q}$. Placing a high logic level on the R

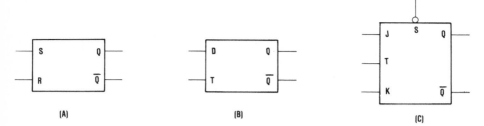

Figure 10-6:   Logic symbols for three basic types of flip-flops. (A) is for an
RS flip-flop, (B) is for a D-type, and (C) is for the JK type

INPUTS		OUTPUTS	
R	S	Q	$\overline{Q}$
HIGH	LOW	LOW	HIGH
LOW	HIGH	HIGH	LOW
LOW	LOW	UNCHANGED	
HIGH	HIGH	NOT PERMITTED	

Figure 10-7:   RS flip-flop logic table. When checking this device, you'll find
that when S input is high and R is low, the flip-flop is set as
shown. When S is low and R is high, you'll have the device set
again, but in an exactly opposite state from that shown. If you
try any other combinations, the circuit will malfunction.

input will reverse the output. However, to do this requires that you
hold the unused input (either S or R) at logic 0 level. All the possible
conditions for the RS flip-flop are shown in Figure 10-7.

The D-type flip-flop is much the same as the RS type just
described except that you'll have to apply low-to-high transition to
the T input for the D input to change state. A troubleshooting logic
table for the D-type is shown in Figure 10-8.

The JK flip-flop symbol (see Figure 10-6) shows that there are
five inputs; S, C, J, K, and T. The outputs are labeled the same as the
RS and D-types, Q and $\overline{Q}$. When checking a JK flip-flop, you'll find
that the S and C inputs are used to preset the device to a particular
state (high or low) before the next operation is started. The T input is
the clock input. You'll often see the S and C inputs referred to as
asynchronous inputs because they do not require a clock pulse on the
T input to change the flip-flop's state before the next operation.
However, the J and K inputs do require a clock pulse on input T
before they can affect the Q and $\overline{Q}$ outputs.

INPUT		OUTPUT	
D	T	Q	$\overline{Q}$
LOW	LOW	PREVIOUS STATE	
LOW	HIGH	LOW	HIGH
HIGH	LOW	PREVIOUS STATE	
HIGH	HIGH	HIGH	LOW

Figure 10-8:   Troubleshooting logic table for a D-type flip-flop

If you inject a logic state into J and the same into K, you'll see a change in state if the clock input (T) goes from a low state to a high state. To set a flip-flop of this type, inject a high-level pulse to the J input and a low-level to the K input. Now, as we have said, you must have a clock input at T, so inject a low-to-high clock pulse into T. Like most flip-flops, the JK can be easily and quickly checked by measuring the voltage states on the outputs, and using a high-to-low, or low-to-high transition to the inputs to toggle the circuits.

A Guide to Digital Pulses

When you're working with digital circuitry, you must deal with steady DC levels, as well as pulses of DC. Figure 10-9 illustrates the transitions used to trigger digital circuits such as flip-flops and other logic elements.

Digital pulses are defined as transitions that occur during a certain period of time. You'll find that timing pulses are generally generated by a clock circuit. Although Figure 10-9 shows a perfect abrupt change in voltage level ... a perfect rectangular waveform ... in actuality, the output of the clock isn't really perfect. Furthermore, the output is normally a set of digital pulses known as a *train of pulses*. However, for simplicity, we will restrict our examination of pulses to one pulse at a time. But, keep in mind that each pulse in the train would be exactly the same, for all practical purposes.

For the pulse shown in Figure 10-10 (and in Figure 10-9), the maximum amplitude is 5 volts. The reference level is 0, or ground. Don't forget, this is only an example and logic high (5V), or logic low (0V, ground) will frequently be other voltage levels.

Referring to the left-hand side of the pulse in Figure 10-10, from 0V to 5V is called the *leading edge*. The right-hand side (5V to 0) is called the *trailing edge*. The width of the pulse that is shaded on the

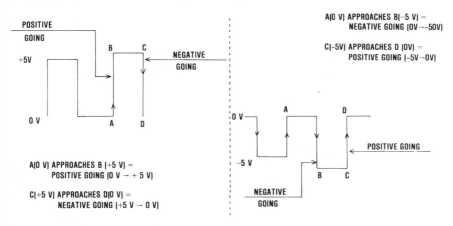

Figure 10-9: Positive and negative-going transitions used to trigger digital circuitry

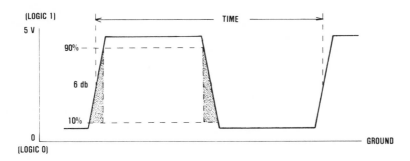

Figure 10-10: Digital pulse terminology. Leading edge, trailing edge, pulse rise time, fall time, etc. See text.

left side (10% to 90%), is known as the *pulse rise time*. The shaded area on the trailing edge (right side 90% to 10%), is the *fall time*. Digital pulses are normally very fast, with very sharp rise and fall times, which is why a very good oscilloscope is usually required to view a pulse or train of pulses.

The time duration is shown at the top of the pulse and is labeled *time*. Notice, the time duration is measured from a reference point of 6dB down (50%) from maximum pulse voltage amplitude, in our example, and includes one entire cycle. The total period (P) of the waveform is equal to 1/F. The duty cycle equals total time divided by the period or, T_{total}/P.

Two other important terms are *Positive Logic* and *Negative Logic*. Figure 10-10 is showing positive logic because in positive logic, the maximum amplitude (+5V, in this case) is defined as logic level 1, and the ground, or 0, is labeled logic 0). On the other hand, negative logic would equate the voltage level +5 as the digital number 0 and the ground, or zero level, as digital 1. Remember, a high or low logic level may be only relative to each other, i.e., one voltage in respect to the other (for example, 0.4V, a low, to 5V, a high).

Practical Input/Output Signal Combinations for AND and OR Gate IC's

Many manufacturers use a small circle to indicate a relatively low (L) level. If there is no small circle on the drawing of the logic symbol input or output(**IC** pins), it indicates the logic level at that point is relatively high (H). Generally, if the small circle is on an input, it means that a relatively low input is required to activate the device. No small circle normally means that the circuit needs a high logic level to activate it. But, if the small circle is on a gate's output, it means you should expect to find an output that is relatively low. Or, as before, no circle on an out means you should find it a high level at that point.

As an example, if you invert the inputs and output of an OR gate, you'll see that the output will be the same as that of an AND gate—both will be positive logic (in other words, 1 is high, 0 is low). Using small circles to indicate lows, Figure 10-11 shows the logic symbols and function table. The OR gate shows L, L, for an input and L for an output. Invert, and you will have H, H, in and an H out.

VARIABLE AND EQUIVALENT		TRUTH TABLE		
AND	OR	A	B	X
		H	H	H
A ⟩— X	A ⟩o— o⟩ X	H	L	L
B	B	L	H	L
		L	L	L

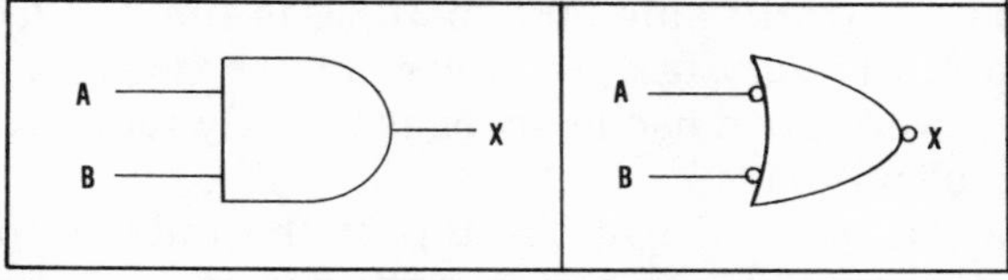

Figure 10-11: Logic symbols and function tables for an OR and AND gate. This shows that inverting the input and output of an OR gate will produce the same output as an AND gate, an important point for the troubleshooter or experimenter/designer. Read the small circle as a relatively low logic level

There are quite a few possible equivalent combinations of the AND and OR gates. Figure 10-12 shows several other equivalent input/outputs that can be achieved by varying the logic levels in the gate connecting pins. For instance, referring to item number 1 in Figure 10-12, you'll notice that a high logic level on both inputs of the AND gate will produce a low logic level on the output (shown by the small circle in the logic diagram of the AND gate, or as an L in the truth table). Looking at the OR gate, you'll see that the output will be a high with two low-level inputs. Invert this level and you will have the same as the AND gate. The rest of Figure 10-12 shows six other combinations and resulting truth tables.

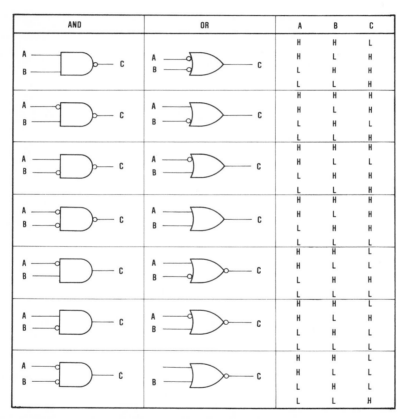

AND	OR	A	B	C
		H	H	L
		H	L	H
		L	H	H
		L	L	H
		H	H	H
		H	L	H
		L	H	L
		L	L	H
		H	H	H
		H	L	L
		L	H	H
		L	L	H
		H	H	H
		H	L	H
		L	H	H
		L	L	L
		H	H	L
		H	L	L
		L	H	H
		L	L	L
		H	H	L
		H	L	H
		L	H	L
		L	L	L
		H	H	L
		H	L	L
		L	H	L
		L	L	H

Figure 10-12: AND and OR gate logic level equivalent combinations that will produce the outputs shown in the truth table alongside the logic symbols

APPENDIX **A**

Reference Tables

MULTIPLICATION FACTOR	PREFIX	SYMBOL	MEANING (USA)	MEANING (FOREIGN COUNTRIES)
$1{,}000{,}000{,}000{,}000{,}000{,}000 = 10^{18}$	exa	E	ONE QUINTILLION TIMES	TRILLION
$1{,}000{,}000{,}000{,}000{,}000 = 10^{15}$	peta	P	ONE QUADRILLION TIMES	THOUSAND BILLION
$1{,}000{,}000{,}00{,}000 = 10^{12}$	tera	T	ONE TRILLION TIMES	BILLION
$1{,}000{,}000{,}000 = 10^{9}$	giga	G	ONE BILLION TIMES	MILLIARD
$1{,}000{,}000 = 10^{6}$	mega	M	ONE MILLION TIMES	
$1{,}000 = 10^{3}$	kilo	k	ONE THOUSAND TIMES	
$100 = 10^{2}$	hecto	h	ONE HUNDRED TIMES	
$10 = 10$	deka	da	TEN TIMES	
$0.1 = 10^{-1}$	deci	d	ONE TENTH OF	
$0.01 = 10^{-2}$	centi	c	ONE HUNDRETH OF	
$0.001 = 10^{-3}$	milli	m	ONE THOUSANDTH OF	
$0.000{,}0001 = 10^{-6}$	micro	μ	ONE MILLIONTH OF	MILLIARDTH
$0.000{,}000{,}001 = 10^{-9}$	nano	n	ONE BILLIONTH OF	BILLIONTH
$0.000{,}000{,}000{,}001 = 10^{-12}$	pico	p	ONE TRILLIONTH OF	THOUSAND BILLIONTH
$0.000{,}000{,}000{,}000{,}001 = 10^{-15}$	femto	f	ONE QUADRILLIONTH OF	TRILLIONTH
$0.000{,}000{,}000{,}000{,}000{,}001 = 10^{-18}$	atto	a	ONE QUINTILLIONTH OF	

Table A-1 Metric Prefixes

METRIC UNITS				
QUANTITY	COMMON UNITS	SYMBOL	ACCEPTABLE EQUIVALENT	SYMBOL
LENGTH	KILOMETER	KM		
	METER	M		
	CENTIMETER	CM		
	MILLIMETER	MM		
	MICROMETER	μM		
AREA	SQUARE KILOMETER	KM^2		
	SQUARE HECTOMETER	HM^2		
	SQUARE METER	M^2	HECTARE	HA
	SQUARE CENTIMETER	CM^2		
	SQUARE MILLIMETER	MM^2		
VOLUME	CUBIC METER	M^3		
	CUBIC DECIMETER	DM^3	LITER	L
	CUBIC CENTIMETER	CM^3	MILLILITER	ML
VELOCITY	METER PER SECOND	M/S		
	KILOMETER PER HOUR	KM/H		
ACCELERATION	METER PER SECOND SQUARED	M/S^2		
FREQUENCY	MEGAHERTZ	MHz		
	KILOHERTZ	kHz		
	HERTZ	Hz		
MASS	MEGAGRAM	MG		
	KILOGRAM	KG	METRIC TON	
	GRAM	G		
	MILLIGRAM	MG		
DENSITY	KILOGRAM PER CUBIC METER	KG/M^3	GRAM PER LITER	G/L
FORCE	KILONEWTON	KN		
	NEWTON	N		
PRESSURE	KILOPASCAL	KPA		
ENERGY, WORK OR QUANTITY OF HEAT	MEGAJOULE	MJ		
	KILOJOULE	KJ		
	JOULE	J		
	KILOWATT-HOUR	KWH		
POWER OR HEAT FLOW RATE	KILOWATT	KW		
	WATT	W		
TEMPERATURE	KELVIN	K		
	DEGREE CELSIUS	°C		
ELECTRIC CURRENT	AMPERE	A		
QUANTITY OF ELECTRICITY	COULOMB	C		
	AMPERE-HOUR	AH		
ELECTROMOTIVE FORCE	VOLT	V		
ELECTRIC RESISTANCE	OHM	Ω		
LUMINOUS INTENSITY	CANDELA			

Table A-2 Metric Units

	IF YOU KNOW	MULTIPLY BY	TO FIND
LENGTH	INCHES	25.4	= MILLIMETERS
	FEET	0.305	= METERS
	YARDS	0.914	= METERS
	MILES	1.609	= KILOMETERS
AREA	SQUARE YARDS	0.836	= SQUARE METERS
	ACRES	0.405	= HECTORES
VOLUME	QUARTS (LG)	0.946	= LITERS
	CUBIC YARDS	0.765	= CUBIC METERS
MASS	OUNCES (AVDP)	28.35	= GRAMS
	POUNDS (AVDP)	0.454	= KILOGRAMS
TEMPER-ATURE	DEGREES FAHRENHEIT	5/9 (AFTER SUB-TRACTING 32)	= DEGREES CELSIUS
LENGTH	MILLIMETERS	0.039	= INCHES
	METERS	3.281	= FEET
	METERS	1.094	= YARDS
	KILOMETERS	0.621	= MILES
AREA	SQUARE METERS	1.196	= SQUARE YARDS
	HECTARES	2.471	= ACRES
VOLUME	LITERS	1.057	= QUARTS (LG)
	CUBIC METERS	1.308	= CUBIC YARDS
MASS	GRAMS	0.035	= OUNCES (AVDP)
	KILOGRAMS	2.205	=POUNDS (AVDP)
TEMPERA-TURE	DEGREES CELSIUS	9/5 THEN ADD 32	= DEGREES FAHRENHEIT

Table A-3 Common Metric Conversion

Glossary of Solid State Terms

and Abbreviations

Active Element ... Any element capable of gain or control. Example: transistor amplifier.

a-d ... Analog to digital.

A/D Converter ... Analog to digital converter. Purpose: to convert analog voltages to their digital equivalents.

Adder ... An arrangement of logic gates that adds two binary digits and produces sum and carry outputs. Purpose: to produce an output that is the sum of its inputs.

Analog IC ... A linear IC. Purpose: to provide an output that is a linear function of the input.

AND Gate ... A gate that produces a logic level 1 on its output, when all inputs are logic level 1. Or, a logic level 0 output if any input is logic level 1, and the others are logic 0. A logic diagram is shown in Figure G-1.

BCD ... Binary coded decimal (see Binary Coded).

BEST ... Base-emitter self-aligned technology.

Binary ... In the binary numbering system, there are two digits (1 and 0).

Binary Coded ... A series of binary digits. Example: a four-bit system (generally called *binary coded decimal* or **BCD**) where the decimal numbers (0 to 9) are converted to binary forms using four binary digits. In a microcomputer, typically an 8-bit code.

Bipolar ... Transistor within an **IC**, or as a discrete unit that utilizes both hole current and electron current when conducting. A bipolar transistor diagram is shown in Figure G-2.

Bit ... A single logic state, 1 or 0; may be called *logic high* or *logic low*.

Buffer ... Amplifier. Purpose: generally used for adapting a circuit to an otherwise incompatible circuit, (see Figure G-3).

Bus ... Several electrical lines being used for a common purpose.

Byte ... The number of bits that a computer processes as a unit. Example: an 8-bit byte.

CCD ... Charge-coupled device. Semiconductor device with cells electrostatically coupled and capable of temporarily storing information in a pattern of pulses.

CDI ... Collector diffusion isolation.

Change State ... To go from a high logic state(1) to a low logic state (0), or the inverse.

Clock ... Normally short for clock pulse generator. Purpose: to produce clock signals for a microcomputer or other digital equipment that requires timing and synchronization. The output clock is usually a series of equally spaced pulses for timing the operation of the circuits.

Closed-Loop Gain ... The actual gain of an operational amplifier when a feedback circuit is connected.

CMOS ... Complementary-metal-oxide semiconductor.

Common Mode ... Signal or test voltages which have equal effect on both inputs of an operational amplifier.

Common Mode Rejection ... A measure of the effective attenuation of a signal that appears simultaneously and in phase at both input terminals of a differential amplifier.

COS/MOS ... Complementary metal oxide insulated gate field effect transistor or **IC**.

Counter ... A series of cascaded flip-flops. Purpose: in microcomputers, counters are used to hold and manipulate binary numbers in electrical (pulse or logic) form. This is a typical use and such systems generally have one stage for each binary bit to be held or manipulated.

D/A Converter ... Digital to analog converter. Purpose: to convert digital number to their analog equivalents.

DAA ... Data access arrangement.

DCTL ... Direct-coupled transistor logic.

Debugging ... The process of finding an undesired *instruction* to a microcomputer or similar device. Do not confuse debugging with *troubleshooting*. Troubleshooting a computer is a term used to mean looking for a fault in the electrical or mechanical system.Debugging usually means finding a fault in a program.

Decoder ... Circuit used to convert a coded or a numbering system from one code to another, such as from hexadecimal to decimal.

Decremental ... Having to do with a counter used to count serial pulses. Example: a program counter in a microcomputer receives one pulse for each step in a program. This counter is *incremented* (or advanced) for each pulse representing a step. When each step or pulse removes one count, the microcomputer's counter is referred to as being decremented.

Differential Comparator ... A differential circuit for indicating when two input signals are equal.

Diode-Resistor Logic (DRL) ... Logic gates such as OR and AND gates, constructed using a combination of diodes and resistors.

DIP ... Dual in-line package (see Figure G-12).

DMA ... Direct memory access.

DTL ... Diode transistor logic.

D-type flip-flop ... A flip-flop for accepting and holding data.

Duty Cycle ... The percentage of time a signal (usually pulsed) or circuit is turned on, or operating at full output. The *duty factor* of such a circuit or signal is usually calculated by multiplying pulses-per-second × pulse width.

ECL ... Emitter-coupled logic. Transistor logic in which more than one input is connected to more than one emitter of an input circuit transistor.

Encoder ... The terms *decoder, encoder,* and *multiplexer* are often interchanged, in technical literature. The term *encoder* is usually applied when an uncoded value is converted to some coded form. Example: converting analog to a binary by use of an A/D converter.

Exclusive-OR Gate ... This will produce an output when all inputs are either logic high or logic low. But not when the inputs are different (see Figure G-14).

Fan In ... This term may have several meanings: (1) the total number of inputs to a digital **IC**; (2) the number of controls in a single digital **IC** that individually, or in combination, produces the same digital output; or (3) the number of inputs to a digital device that are more than normally provided.

Fan-Out ... An **IC** or some other digital circuit that has a single output mode that feeds more than one (generally several in parallel) loads.

Firmware ... Typically, manufacturers use this term to describe something between hardware (an **IC, ROM,** etc.) and software (programs, instructions, and so on).

Flag ... The term *flag* refers to a signal that can be used for any number of functions. Example: a flag could be used to halt a microcomputer at a certain number of steps. You'll find that the designation *flag* can be applied to any signal that indicates that a particular condition in digital logic has occurred (full count, error, etc.).

Floppy Disk ... A removable magnetic storage medium that is permanently contained in a paper envelope. Purpose: most commonly used as a magnetic storage device for microcomputers.

Full Adder ... An adder which accepts two inputs, as in a half-adder, and produces an output, but (in addition) accepts a third input (a carry input).

Gate Expander ... An expandable circuit used for several inputs to a single gate.

Handshake Bus ... A bus that interconnects a microprocessor-based system to the outside world.

Hardware ... The actual components (electrical and mechanical) that are used to build a given system. Example: in logic circuitry, a Random Access Memory, Read Only Memory, Integrated Circuits, Flip-Flops, and so on.

HMOS ... High performance (see *n-MOS*).

Hex ... This term is short for hexadecimal. In this numbering system, each binary digit represents four binary digits.

Hybrid ... Strictly, a mixture or combination of two different technologies. Example: an amplifier that employs both discrete and integrated circuits or a computer that has both analog and digital capabilities. However, the term *hybrid* can be used with any combination of basically different components, materials, or arrangement of circuits.

I^{2}L ... Integrated injection logic. Applications: microprocessors, watches, counters, timers and all sorts of equipment such as autos, etc.

IMM ... Integrated magnetic memory.

Inhibit ... To prevent the operation of a certain circuit such as a gate, etc. This operational stoppage is usually done by a blocking pulse or other form of blocking signal.

Input Current Offset ... The effective input current (usually not zero) drain by the unbalanced input circuits of an operational amplifier.

Input Voltage Offset ... The effective input voltage (usually not zero) caused by an unbalance in input circuits of an operational amplifier.

Insulated Gate Field Effect Transistor Logic (MOS/FET) ... Logic employing insulated gate field effect transistors. A standard **MOS/FET** schematic symbol is shown in Figure G-4.

Interface Circuit ... The hardware for linking two units of electronic equipment. Example: hardware components to link a microcomputer with its inputs or outputs. For instance, peripheral data bus lines are usually bidirectional and capable of transferring data to and from the microprocessor and memory within a microcomputer.

Inverter ... An **IC** whose function is simply to invert. Example: change a logic low (0) to a logic high (1). Typically, a small circle is used on the inverted output or input line and the inversion is in respect to another **IC** pin, etc. (see Figure G-5).

Inverting Input ... In an operational amplifier, the input terminal that will produce an inverted signal on the output. You'll find that the degenerative feedback path is from the output back to the inverting input pin of the operational amplifier.

ISL ... Injection Schottky logic.

JFET ... Junction field effect transistor or **IC**. The gate is uninsulated from the channel and draws current. See Figure G-6 for a standard **JFET** schematic symbol.

JK Flip-Flop ... A flip-flop meeting the Boolean algebra equation $Q+ = \overline{Q}J + Q\overline{K}$. It counts clock input pulses when JK = 1. The R and C inputs preset the circuit to a desired state before another operation is begun. See Figure G-7 for a JK flip-flop schematic symbol.

Karnaugh Map ... An illustration (a truth table) drawn in such a way that it simplifies a Boolean expression.

Latch ... Typically, a feedback circuit used in a flip-flop which, when pulsed, assumes and retains a certain state. Basically, it is a simple logic storage element and is frequently an R-S type flip-flop

The latch is controlled by two input steering lines called *set* (S) and *reset* (R).

Linear Mode ... When any circuit or component is operated in such a manner that its output is a linear function of its input.

Locking Range ... This phrase is used most often in reference to the frequency range over which a phase-locked loop will follow an input signal without losing lock.

Logic Circuit ... Any circuit that provides an input-output relationship corresponding to a Boolean algebra logic function. Example: JK flip-flops.

Logic 1 ... One of two possible values of a binary digit (see *Logic 0* for the other possibility). If the pulse is a voltage, logic 1 is some value greater than logic 0. Example: 5 volts.

Logic State ... In a binary system, high voltage level might mean a 1, and low voltage level might mean a 0. Logic state means the high or low *state* of a circuit.

Logic 0 ... One of two possible values of a binary digit (see *Logic 1* for the other possibility). If the pulse is a voltage, logic 0 is some value lower than logic 1. Example: 0.4 volts.

LSI ... Large scale integration.

MESFET ... Metal semiconductor field effect transistor

MIL Specifications ... Military standards specifications. Example: MIL-STD-806B. This specification says that a small circle drawn on the input of a logic element indicates that a relatively low input signal activates the function. *Note:* this is by no means the only specification given in these standards. It is only an example of what MIL Specifications means.

MOSFET ... Metallic oxide insulated gate field effect transistor. See Figure G-4 for a standard **MOS/FET** schematic symbol.

MSI ... Medium scale integration

NAND Gate ... A gate (see Figure G-8) that produces a logic low when, and only when, all inputs are at a logic high. Non-coincident input pulses have no effect on the output of a NAND gate. The output always stays at logic 1 level.

N-Channel FET ... An **FET** (or **IC**) in which the channel between the source and drain is doped with N-type semiconductor material.

Negative Logic ... Inverted position logic, i.e., the logic high state (1) is *more negative* than the logic low (0) state.

Non-Inverting Input ... When speaking of an operational amplifier (which is generally the case), it is the input pin that produces an in-phase signal on the output pin.

NOR Gate ... A gate that produces a logic low (0) output when either, or both, inputs are at a logic high (1). The Boolean equation is A • B = C. See Figure G-9 for the standard schematic symbol for a NOR gate.

n-MOS ... N-channel metal oxide semiconductor. This logic family has several different names: **HMOS, X-MOS, S-MOS.**

Open-Loop Gain ... This term is usually associated with the gain of an operational amplifier. Example: the open-loop voltage gain is a ratio of output signal voltage to the differential input signal voltage of the operational amplifier, measured with no feedback being applied.

Operational Amplifier (OP AMP) ... A high-gain DC amplifier that depends on negative feedback for a precise gain. Generally, a linear **IC** using external components to control an external feedback from the output to the input, to determine the **OP AMP's** functional characteristics. See Figure G-10 for a standard **OP AMP** schematic symbol.

OR Gate ... A gate that will produce a logic high (1) when either, or both, inputs are at a logic high. Boolean equation is A + B = C. See Figure G-11 for an OR gate schematic symbol.

Positive Logic ... The opposite of negative logic, i.e., logic in which the logic high (1) state is more positive than the logic low (0) state.

Programmable Read Only Memory (PROM) ... An **IC** that is shipped by the manufacturer with all its bits at each address blank (or at a binary logic low (0) level, in most cases). Generally, the user must program each bit at each address by means of a current. Also, once the **PROM** is programmed, it is *permanent* and *irreversible*. This includes any mistake that one may make. In other words, the entire **PROM** is useless if a mistake is made during programming. Two solutions to this problem are: (1) use automatic or semi-automatic **PROM** programming equipment or, (2) there are **IC's** called **EAROM's** (electrically alterable read only memory) that can be erased and reprogrammed.

Random Access Memory (RAM) ... An **IC** memory used in computers, etc., to store temporary data (the information to be acted upon by the computer program). Example: Motorola's 128 by 8 static **RAM**. This **IC's** matrix has 128 addresses, each with 8 bits. Thus, 8-bit bytes can be read into, or out of, 128 locations.

Read Only Memory (ROM) ... An **ROM** is used in a computer to store the computer program or instructions and is usually in a separate **IC** package. Example: Motorola's 1024 by 8 **ROM**. In this **IC**, there are 1024 addresses, each with 8 bits. Thus, 8-bit binary words or bytes can be read out of 1024 locations.

Register ... Generally, a series of cascaded flip-flops used to hold and manipulate binary numbers (computer word or byte) in electrical (pulse or logic level) form. Purpose: a register is similar to a counter (used to count each step of a microcomputer program) except that the primary function of a register is to hold the binary numbers until they are needed at some later time. Example: hold a binary number in temporary storage so that it may be added to another number in memory.

Resistor-Transistor Logic (RTL) ... Logic gates formed using combinations of resistors and transistors.

Ring Counter ... A series of flip-flops with the output returned to the input. As input signals are counted, a certain state moves around the loop in an ordered sequence. Purpose: used to store several bits of information. Also, will accept shift instructions that cause all stored information to shift one position at a time (left or right).

Ripple Counter ... A simple serial counter without feedback is known as a ripple counter. In this type circuit, pulses enter the counter and, if the counter fills or reaches some other desired level, a flag or signal may be sent to other circuits, indicating a full count. At this point, the counter usually starts counting all over again. The ring counter is much like a shift register in that clock pulses are fed to all stages. However, unlike the ripple counter, a shift register does use feedback.

R-S Flip-Flop ... This circuit uses inputs labeled S and R. The outputs are Q and $\overline{Q}$. The Boolean algebra equation is Q+ = S + Q$\overline{R}$, with the restriction RS = 0.

Seven-Segment Display ... A display unit constructed by using seven bars. The arrangement of the bars permits each digit from 0 to 9 to be displayed on the face of the unit. See Figure G-13.

Shift Register ... In microcomputers, these circuits (usually flip-flops) are used to hold and manipulate binary numbers in electrical form (pulse or logic level). Example: in a microcomputer, an 8-bit register would have eight flip-flops. In a shift register of this type, the contents may be shifted by one position to the right or left by a shift signal applied to all stages simultaneously. As a result of this shift, a binary number can be changed to a higher or lower value. For example, from a decimal 8 to 16, etc.

Slew Rate ... Basically, the maximum rate of change of an amplifier (such as an **OP AMP**) output voltage, usually measured under certain conditions. The maximum rate of output voltage change is measured under closed loop conditions and with large signal conditions (when an AC input voltage causes the amplifier to go into saturation).

S-MOS ... See *n-MOS*.

SOS ... Silicon on Sapphire.

SSI ... Small scale integration.

STL ... Schottky transistor logic.

Toggle ... To turn off and on. Example: A flip-flop is sometimes called a toggle because its operation is very similar to that of a mechanical toggle switch in that it is either off or on.

TTL ... Transistor-transistor logic.

VFET ... Vertical field effect transistor.

VLSI ... Very large scale integration.

Z-MOS ... See *n-MOS*.

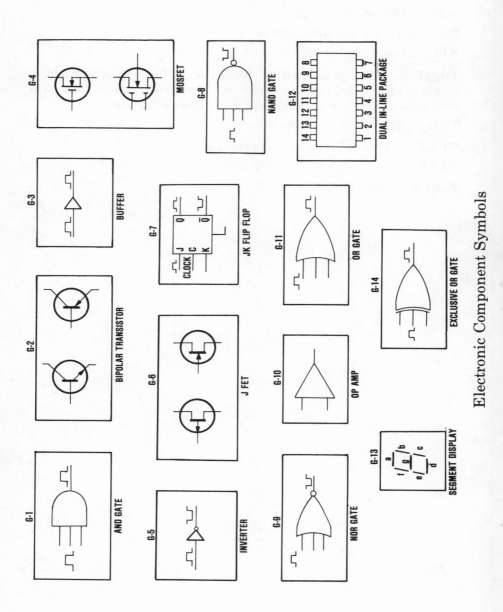

Electronic Component Symbols

232

INDEX